Internal Assessment for
Chemistry

Skills for success

Christopher Talbot

HODDER
EDUCATION

Author profile

Christopher Talbot graduated with honours in biochemistry from the University of Sussex in the United Kingdom. He also studied molecular biology in the crystallography department at Birkbeck College in London. He has a master's degree in life sciences (chemistry) from the National Technological University in the Republic of Singapore. He has taught IB chemistry, IB biology and theory of knowledge at the Anglo-Chinese School (Independent) and the Overseas Family School in Singapore.

Author's acknowledgements

I am grateful to my former colleague, Dr David Fairley, Overseas Family School, Singapore for his rigorous editing and reviewing of early drafts of all the chapters and his introduction to Python programming. I am also indebted to Dr Caroline Evans, Head of Chemistry, Wellington College, United Kingdom for her guidance and advice on the chapters focused on the group 4 assessment criteria. It was also a pleasure to work with Dr Andrew Davies, Head of Environmental Studies, St Edward's School, United Kingdom and author of *Internal Assessment for Biology for the IB Diploma: Skills for Success*. Dr John Green gave advice on very early drafts of the material. Professor Laurence Harwood, Reading University, and Professor Jim Hanson, Sussex University advised on Chapter 3: Preparative organic chemistry.

Photo credits

The Publishers would like to thank the following for permission to reproduce copyright material.

pp.vi–vii © Samart Boonyang/123rf; **p.xi** Adapted with permission from Talanquer, V., *Central Ideas in Chemistry: an alternative perspective*, JCE, 2016 93, 3–8. Copyright (2016) American Chemical Society; **p.54** Screenshot reproduced by kind permission of Professor Craig Merlic, UCLA Department of Chemistry & Biochemistry; **p.55** Screenshot of ChemKinetics reproduced by kind permission of Professor Stephen W. Bigger, Victoria University; **p.56** Screenshot reproduced by kind permission of Scott A. Sinex, Professor Emeritus, Prince George's Community College; **p.73** © picsfive/Adobe Stock

Every effort has been made to trace all copyright holders, but if any have been inadvertently overlooked, the Publishers will be pleased to make the necessary arrangements at the first opportunity.

Although every effort has been made to ensure that website addresses are correct at time of going to press, Hodder Education cannot be held responsible for the content of any website mentioned in this book. It is sometimes possible to find a relocated web page by typing in the address of the home page for a website in the URL window of your browser.

Hachette UK's policy is to use papers that are natural, renewable and recyclable products and made from wood grown in sustainable forests. The logging and manufacturing processes are expected to conform to the environmental regulations of the country of origin.

Orders: please contact Bookpoint Ltd, 130 Park Drive, Milton Park, Abingdon, Oxon OX14 4SE. Telephone: (44) 01235 827827. Fax: (44) 01235 400401. Email: education@bookpoint.co.uk

Lines are open from 9 a.m. to 5 p.m., Monday to Saturday, with a 24-hour message answering service. You can also order through our website: www.hoddereducation.com

Contents

Introduction

How to use this book

There are two aspects to the practical work in the IB Chemistry programme:

1. general practical work
2. a single individual investigation – an internal assessment project of 10 hours duration.

This publication is aimed specifically at IB Chemistry students and is to be used throughout your two years of study. Practical activities and the internal assessment form an essential part of the IB Chemistry syllabus (first assessment held in 2016) with 40 hours recommended teaching time for standard level and 60 hours for higher level. This represents an average 25 % of the total teaching time. The internal assessment is worth 20 % of the final assessment.

General practical work includes experiments in the laboratory, spreadsheet or online **simulations** (for example, Java Applets, Flash animations or Python (Trinket) simulations), demonstrations by your teacher and class activities of a formative nature. These are designed to help you learn chemistry via practical work.

The 'Applications and skills' section of the IB Chemistry syllabus lists specific laboratory skills, techniques and experiments that you must experience at some point during your study of the IB Chemistry course. Your school is likely to arrange additional practical work covering other topics in the IB Chemistry programme. It is the skills and not the specific experiments that will be assessed in the written examinations. Other recommended laboratory skills, techniques and experiments are listed in the 'Aims' section of the IB Chemistry syllabus.

Within the IB Chemistry syllabus, there is also a specific set of mandatory practicals that you will carry out over the course and your knowledge and understanding of these will be assessed in your final examination papers.

This guide will ensure you can aim for your best grade by:

■ building practical, mathematical and analytical skills for the mandatory and other common practicals through a comprehensive range of strategies and detailed examiner advice and expert tips

■ offering concise, clear explanations of all the IB requirements, such as the assessment objectives of each criterion for the internal assessment, including checklists and rules on academic honesty

■ demonstrating what is required to obtain the best internal assessment grade for the individual investigation with advice and tips, including common mistakes to avoid

■ suggesting practicals that might, if modified, form the basis of an individual investigation

■ making explicit reference to the IB learner profile and the associated Approaches to Learning (ATL) that are central to the IB programme, with their connections to practical work

■ giving worked examples and commentary throughout so you can see the application of physical and mathematical principles and concepts

■ testing your comprehension of the skills covered with embedded activity questions.

Use the space provided in the margins of the book to make your own notes and record your own observations as you progress through the course.

Features of this book

Key definition

The definitions of essential key terms are provided on the page where they appear. These are words that you can be expected to know for exams and practical work. A glossary of other essential terms, highlighted throughout the text, is given at the end of the book.

Examiner guidance

These tips give you advice that is likely to be in line with IB examiners.

Worked example

Some practical skills require you to carry out mathematical calculations, plot graphs, and so on. These examples show you how.

■ ACTIVITY

Questions and suggested outline of possible practice activities.

Ideas for investigations

Ideas for possible investigations.

RESOURCES

Useful websites or published books.

Expert tip

These tips give practical advice that will help you boost your final grade.

Common mistake

These identify typical mistakes that candidates make and explain how you can avoid them.

Studying IB Chemistry

IB Learner Profile

The IB Chemistry course is linked to the IB learner profile. Throughout the course, and while carrying out your internal assessment, you will have the opportunity to develop each aspect of the learner profile: Inquirers, Knowledgeable, Thinkers, Communicators, Principled, Open-minded, Caring, Risk-takers, Balanced and Reflective.

Practicals

Carrying out practicals throughout your IB Chemistry course will give you the opportunity to practice carrying out an investigation, and will give you the scientific skills you need for your internal assessment.

Chemistry

The IB Chemistry course, and the internal assessment in particular, give you the chance to develop the approaches to learning skills:
- thinking skills when planning investigations, collecting data and analyzing your results
- social skills when working with your peers
- communication skills when reporting and presenting your findings
- self-management skills when working independently
- research skills to help plan your investigation, and to put it into context.

Internal Assessment

The internal assessment gives you the opportunity to display the skills and knowledge you have learned throughout your course, while exploring an area of Chemistry that interests you personally.

Studying IB Chemistry

Chemistry and the scientific method

Chemistry is an observational and experimental science. Systematic observations and reliable measurements of chemical phenomena may suggest hypotheses. These hypotheses can lead to experiments which systematically manipulate a **variable** (under controlled conditions) in order to establish a relationship between two variables. This in turn can result in an improved understanding of chemistry. The **scientific method** can be seen as a cycle (Figure 1 and Table 1). Exploration of one chemical phenomenon can lead to further modification, through analysis and **evaluation**, resulting in the **investigation** and **testing** of further hypotheses.

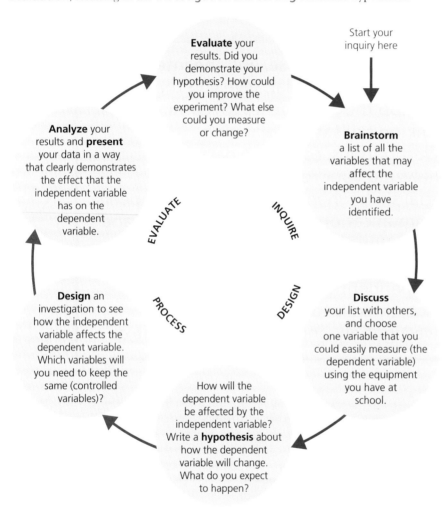

Figure 1 The investigation cycle

Key definitions

Investigation – a scientific study consisting of a controlled experiment in the chemistry laboratory.

Scientific method – the use of controlled observations and measurements during an experiment to test a hypothesis.

Stage of cycle	Description	Key word definitions
1	Formulate your research question: which usually inquires how one variable (the independent variable) affects another (the dependent variable).	Variable – a factor that is being changed, measured, or kept the same during an investigation. Independent variable – the variable that you systematically change (across a range) in an investigation. Dependent variable – the variable that you measure in an investigation. Its value depends on the independent variable. Processed variable – a variable that can be produced by transforming a measured variable through mathematical manipulation.
2	The research question may lead you to formulate a hypothesis. When making a hypothesis, the investigator proposes and explains (via a scientific model) how the independent variable may affect the dependent variable (in a causal relationship). A prediction, ideally quantitative, may be made.	Hypothesis – a tentative explanation based on a scientific model of the observed chemical phenomenon you are investigating using the scientific method. Prediction/predict – predictions are derived from a hypothesis and describe the results you expect to obtain from an investigation.
3	Identify how you can establish a correlation (and ideally a quantitative mathematical relationship) between the two variables (independent and dependent), while controlling other variables.	Controlled variable – a variable that is kept the same in an investigation. In an experiment, at least three controlled variables should be listed, and information about how they will be kept the same included.
4	Develop a method and outline it clearly. You should include a clearly labelled cross-sectional diagram showing the relative positions of the different apparatus and instruments. You should manipulate the independent variable and measure the dependent variable, with other variables controlled (and monitored). You should write your method with sufficient detail and clarity that someone else can follow your instructions and obtain reproducible results.	Control – an experiment where the independent variable is either kept constant or removed. This can be used for comparison, to prove that any changes in the dependent variable in experiments when the independent variable is manipulated must be due to the independent variable rather than other factors.
5	Carry out your investigation and gather raw data. Record your data by measuring the dependent variable. Present your raw data in tables with their appropriate units (typically SI base units) and errors. Mathematically process your raw data in some way (for example, by calculating means, squares, reciprocals or logarithms). Plot a graph (typically a line graph) to present the trends clearly and facilitate mathematical analysis (via the gradient, intercept, area and so on). Transform any non-linear relationships into a linear graph with a line of best fit, where possible and appropriate. You may also record relevant qualitative data (observations) (see Key definitions on page x).	Data – recorded observations and numerical measurements using apparatus and instruments. Measuring/measure – obtain a measured value for a quantity. Qualitative data – observations not involving measurements; for example, colour changes during an experiment, the release of gases (colour and odour) and decrepitation (crackling noises) during heating.
6	Develop an explanation for the results in your analysis. What does the raw, processed and displayed data show you about the relationship between the variables? Analyze the results (data) with reference to the research question and scientific model. Do the results support the hypothesis, or falsify it?	Analysis/analyze – recognize and comment on trends in raw and processed data and state valid conclusions. Explanation/explain – give a detailed account including reasons or causes.
7	Evaluate the investigation and suggest improvements and extensions. When commenting on limitations, consider the procedures, the instruments, the use of instruments, the quality of the data (for example, its accuracy, sensitivity and precision) and the relevance and reliability of the data (and its errors). To what extent might the limitations have affected the results? Propose realistic improvements that address the limitations and increase accuracy or precision.	Evaluated/evaluation – an assessment of the reliability and precision of the raw data recorded during an experiment, the other limitations of the techniques, and the conclusions. Accuracy – measurements or results that are in close agreement with the true or accepted values. Precision – precise measurements are ones in which there is very little spread about the mean value. Sensitivity – the smallest change that can be detected by a measurement by an instrument.
8	The improvements and extensions to the method can lead to further investigations, and so the cycle (of the scientific method) repeats itself.	

Table 1 The scientific method cycle

Replicates can improve the reliability of the raw data generated by an investigation and enable anomalies to be identified. **Repeated** measurements also allow you to reduce the impact of **random error** by averaging.

Raw data refers to data collected without any processing. It is just the values of each variable collected. It is often difficult to use for data analysis, and usually needs to be processed in some way. **Processed data** refers to data that are ready for analysis. Processing implies transforming the data (such as carrying out a calculation).

A processed variable is a variable that is calculated from measured data (from the dependent variable); for example, in a practical to determine the average rate of reaction, the reciprocal of the total time for the reaction to go to completion or reach a certain point is calculated.

> **Key definitions**
>
> **Replicate** – repetition of the entire experiment run at the same time to record repeat measurements and observations.
>
> **Quantitative data** – numerical data (with units and random uncertainty) from measurements, such as the pH during the determination of a titration curve (where volume of titrant is the independent variable).

The nature of chemistry

Chemistry is the study of matter, its properties, how and why particles of substances chemically combine or separate to form new substances, and how substances interact with energy. The borders with physics and biology are 'grey' and it is often known as the 'central science'. Chemical knowledge is both a theoretical and practical subject supported by mathematics and physics.

There are many links between the IB Chemistry topics. For example, stoichiometric relationships underpin many IB Chemistry topics, including energetics/thermochemistry, chemical kinetics and equilibrium. Organic chemistry is underpinned by all of the IB Chemistry topics.

An alternative to topics is to consider the central ideas or concepts that allow understanding of events and phenomena in chemistry across topics. Concepts are used to predict and explain chemical phenomena. One possible summary of chemical concepts is illustrated in Table 2.

Matter is composed of particles (atoms, molecules or ions).
Elements are classified into groups in the periodic table.
Bonds form between atoms by sharing valence electron pairs or by electron transfer or delocalization.
Molecular shape is important in determining properties.
Molecules interact with each other via weak forces.
Energy and mass are conserved.
Energy and matter tend to disperse (entropy).
There are barriers to reaction (activation energy).
There are a few fundamental types of reaction (for example, redox and acid base and ionic precipitation).

Table 2 One possible set of central ideas or concepts in chemistry

Framework for Chemistry

The IB Chemistry syllabus is comprehensive with a large number of interrelated topics and concepts. It helps to simplify the content by using a 'concept map' (Figure 2) to outline the essential components and concepts of a chemistry course and show how they interrelate. The practical skills you will learn during the course can be framed in the context of this diagram. Subsequent chapters will cover essential skills that can be applied to different areas of the syllabus. Figure 2 can be used to help you decide which area of chemistry you want to address in your internal assessment project (Chapter 6).

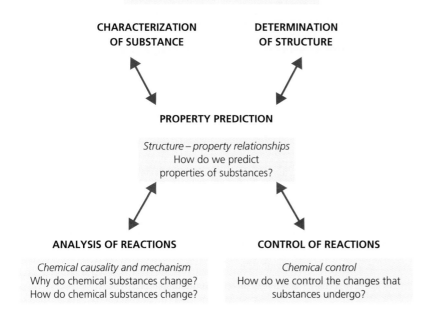

Figure 2 Framework of central concepts for the practice of chemistry (Adapted from Talanquer, V., *Central Ideas in Chemistry: an alternative perspective*, JCE, 93, 3–8)

Chemists analyze and change the material world by modelling it at two different levels: the macroscopic level of bulk materials and a submicroscopic level (atoms, ions and molecules) beneath the macroscopic material world.

IB Chemistry practicals

There are numerous practicals listed in the *IB Chemistry Guide*. This book focuses on practical experiments that can help you with exam questions (Paper 3, section A), and also help you to select and implement a suitable investigation for your internal assessment. The practicals in the guide can be divided into three categories.

■ Mandatory practicals: these are prescribed in the *IB Chemistry Guide*. You need an understanding of these experiments as they may be examined in Paper 3.

■ Useful additional practicals suggested in the guide to be chosen at the discretion of your chemistry teacher. These will not be examined specifically, but may provide useful ideas to help you select and implement your internal assessment project.

■ Computer simulations: information communication technology is encouraged throughout all aspects of your course. Certain skills involving information communication technology are specified in the *IB Chemistry Guide*: these use computers to model or draw associations between numerical data.

Examiner guidance

Secondary data refers to data obtained from another source, such as via reference material or published third-party results. The source of secondary data must always be cited in a report; analysis of secondary data can also form an important part of experimental work. For example, the **NIST Chemistry WebBook** (http://webbook.nist.gov/chemistry/name-ser/) has a wide range of thermodynamic and other data, including spectral data, for analysis or reference.

Mandatory practicals

Subtopic	Mandatory practical
Topic 1.2	Obtaining and using experimental data to derive empirical formulas from reactions involving mass changes (Chapter 2).
Topic 1.3	Using the experimental method of titration to calculate the concentration of a solution by reference to a standard solution (Chapter 2).
Topic 1.3	Obtaining and using experimental values to calculate the molar mass of a gas from the ideal gas equation (Chapter 1).
Topic 5.1	A calorimetry experiment for an enthalpy of reaction should be covered and the results evaluated (Chapter 3).
Topic 8.2	Candidates should have experience of acid–base titrations with different indicators (Chapter 1).
Topic 8.3	Students should be familiar with the use of a pH meter and universal indicator (Chapter 1).
Topic 9.2	Performing laboratory experiments involving a typical voltaic cell using two metal/metal ion half-cells (Chapter 1).
Topic 10.1	Constructing 3D models (real or virtual) of organic molecules (Chapter 3).
Topics 15.1 and 19.1	Performing lab experiments that could include single replacement reactions in aqueous solutions (Chapter 1).

Table 3 List of mandatory practicals for chemistry

Suggested practicals

The following suggested practicals will help enhance your understanding of chemistry and could act as possible individual investigations. Your chemistry teacher may select them to include in your Practical Scheme of Work (PSOW). They are listed in the 'guidance' section of the syllabus under 'Aims' and so will not be examined.

Subtopic	Suggested practical
1.2 Mole concept	Determine the percent by mass of water present in hydrates. Determine the empirical formula of magnesium oxide by combustion. Calculate Avogadro's number.
1.3 Reacting masses and volumes	Gravimetric determination by precipitation of an insoluble salt. Use data-loggers to measure temperature, pressure and volume changes in reactions. Determine the value of the gas constant, R.
2.2 Electron configuration	Observe emission spectra from discharge tubes of different gases (at low pressure). Study spectra from flame tests.
3.2 Periodic trends	Investigate the use of transition metals as catalysts. Identify trends in chlorides or oxides across period 3.
4.1 Chemical bonding and structure	Investigate compounds based on their bond type. Obtain sodium chloride by evaporation.
5.1 Measuring energy changes	Calculate the energy content of food. Determine the enthalpy of melting of ice. Determine the enthalpy change of simple reactions in aqueous solution.
5.2 Hess's law	Investigations based on Hess's law.
5.3 Bond enthalpies	Determine the enthalpy of combustion of propane or butane.
6.1 Collision theory and rates of reaction	Investigate the rate of a reaction with and without a catalyst. Investigate rates by changing the concentration of a reactant or temperature.
7.1 Equilibrium	Investigate chemical systems qualitatively by looking at the effects of pressure, concentration and temperature changes on different equilibrium systems.
8.2 Properties of acids and bases	Investigate the properties of acids and bases.

8.3 pH scale	Perform an acid–base titration by monitoring with an indicator or a pH probe.
8.4 Strong and weak acids	Investigate the strengths of a range of acids and bases.
8.5 Acid deposition	Investigate quantitatively the effects of acid rain on different construction materials.
9.1 Oxidation and reduction	Investigate corrosion and galvanization. Redox titrations including the Winkler method to measure biochemical oxygen demand. Demonstrate the activity series.
9.2 Electrochemical cells	Construct and investigate simple voltaic cells using two metal/metal ion half-cells. Investigate electrolysis of molten salts.
10.1 Fundamentals of organic chemistry	Use distillation to separate organic liquids or use a rotary evaporator to remove a solvent from a mixture.
10.2 Functional group chemistry	Experiments could include distinguishing between alkanes and alkenes, preparing soap and the use of gravity filtration, filtration under vacuum (using a Büchner flask), purification including recrystallization, reflux and distillation, melting point determination and extraction.
13.1 First-row d-block elements	Investigate the oxidation states of vanadium and manganese. Analysis of transition metals via redox titrations.
13.2 Coloured complexes	Investigate the colours of a range of complex ions of elements such as chromium, iron, cobalt, nickel and copper via spectrophotometry.
15.1 Energy cycles	Determine either the enthalpy of crystallization of water or the heat capacity of water when a cube of ice is added to hot water.
16.2 Activation energy	Determine the activation energy for a reaction via an Arrhenius plot.
17.1 Equilibrium	Determine the equilibrium constant for an esterification reaction.
18.1 Lewis acids and bases	Experimental investigation of transition metal complexes.
18.2 Calculations involving acids and bases	Investigate the properties of strong and weak acids.
18.3 pH curves	Investigate pH curves. Determine the pK_a of a weak acid. Prepare and investigate a buffer solution. Determine the pK_a of an indicator. Determine the equivalence point via a conductivity or temperature probe.
20.1 Types of organic reaction	Synthesize a drug or medicine (for example, aspirin) or investigate a household product (for example, fading of tomato ketchup with bromine).
20.3 Stereoisomerism	Synthesize and characterize an enantiomer (for example, (–)menthol) or investigate the resolution of a racemic mixture.
A1 Materials science introduction	Investigate the stretching of rubber bands under different chemical environments. Investigate the properties of metals, polymers, ceramics or composites. Make thin concrete slabs from various ratios of cement, gravel, and sand and investigate their breaking strength upon drying.
A2 Metals and ICP spectroscopy	Calculate the Faraday constant via electrolysis of aqueous copper sulfate. Solve for the concentration of a nickel or copper solution using Beer's law and spectrophotometry. Analysis of alloy composition, such as using colorimetric analysis to determine manganese in a paper clip or gravimetric analysis to determine the silver or copper in a coin.
A3 Catalysts	Investigate the decomposition of potassium sodium tartrate with cobalt chloride. Investigate the decomposition of hydrogen peroxide with manganese(IV) oxide and ion exchange.
A4 Liquid crystals	Investigate a thermotropic liquid crystal and the temperature range which affects these crystals.
A5 Polymers	Investigate the physical properties of high and low density polyethene. Synthesize a polyester, polyamide or other polymer, performing this quantitatively to measure atom efficiency.
A9 Condensation polymers	Synthesize nylon.
A10 Environmental impact of heavy metals	Investigate waste water treatment.

Table 4 Suggested practicals

Computer simulations

Other practical skills involve the use of information communication technology (Table 5).

Subtopic	Activity/simulation
2.1 The nuclear atom	Simulation of **Rutherford's gold foil experiment**. (http://micro.magnet.fsu.edu/electromag/java/rutherford/)
4.1 Ionic bonding and structure	Computer simulation to observe **crystal lattice structures**. (http://www.chemtube3d.com/solidstate/_simplecubic(final).htm)
4.3 Covalent structures	Computer simulation to model **VSEPR structures**. (http://www.chemtube3d.com/VSEPRShapeH2O.html)
4.4 Intermolecular forces	Computer simulations of **intermolecular forces interactions**. (http://www.chemtube3d.com/ElectrostaticSurfacesPolar.html)
4.5 Metallic bonding	Computer simulations to show examples of **metallic bonding**. (https://www.pbslearningmedia.org/resource/phy03.sci.phys.matter.metal/the-structure-of-metal/#.Wj9dGa2cYnU)
6.1 Collision theory and rates of reaction	Use of simulations to show how **molecular collisions** are affected by change of macroscopic properties such as temperature, pressure and concentration. (https://phet.colorado.edu/en/simulation/legacy/reactions-and-rates)
7.1 Equilibrium	Use of simulations and animations to illustrate the concept of **dynamic equilibrium**. (https://phet.colorado.edu/en/simulation/legacy/reversible-reactions)
12.1 Electrons in atoms	Simulations of the **Davisson–Germer electron diffraction experiment**. (https://phet.colorado.edu/en/simulation/davisson-germer)
14.2 Hybridization	Computer simulations of **hybrid orbitals**. (https://www.youtube.com/watch?v=SJdllffWUqg)
16.2 Activation energy	Simulations and virtual experiments to study the **effects of temperature and steric factors** on rates of reaction. (http://michele.usc.edu/105b/kinetics/brhinew2.html)
18.3 pH curves	Simulations of **titration curves**. (http://users.wfu.edu/ylwong/chem/titrationsimulator/index.html)
A2 Metals and ICP	Simulations of semiconductors. (https://phet.colorado.edu/en/simulation/legacy/semiconductor)
A3 Catalysts	Simulation of nanoparticles as catalysts. (http://www.chemtube3d.com/AuNano_home.html)
A4 Liquid crystals	Computer animations of thermotropic liquid crystals. (https://www.doitpoms.ac.uk/tlplib/liquid_crystals/printall.php)
A6 Nanotechnology	Simulation of carbon nanotubes. (https://www.youtube.com/watch?v=R4m54nYPjPQ)
A8 Superconducting metals and crystallography	Simulations of superconducting metals (https://www.youtube.com/watch?v=vANjWS8YhgU) and crystallography (http://escher.epfl.ch/Bragg/).
B2 Proteins and enzymes	Simulation of gel electrophoresis. (http://learn.genetics.utah.edu/content/labs/gel/)
C3 Nuclear fusion and fission	Simulations of radioactive decay. (https://phet.colorado.edu/en/simulation/legacy/alpha-decay) Simulations of nuclear fission and fusion. (https://phet.colorado.edu/en/simulation/nuclear-fission) (http://hypnagogic.net/sim/Sim/fusion/Fusion.html)
D3 Opiates	Computer animations for the investigation of 3D visualization of drugs and receptors. (https://pymol.org/2/)

Table 5 Computer activities

Approaches to learning

Approaches to learning (ATL) are deliberate strategies, skills and attitudes that underlie all aspects of the IB diploma programme. These approaches are intrinsically linked with the IB learner profile attributes (see page xvi), and are designed to enhance your learning and preparation for the IB Diploma Programme assessment and beyond.

Expert tip

ATLs encompass the key values and principles that underpin an IB education.

The aims of ATLs in the IB Diploma Programme are to:

■ link prior knowledge to course-specific understanding, and make connections between different subjects

■ encourage you to develop a variety of skills that will equip you to continue to be actively engaged in learning after you leave your school or college

■ help you not only to obtain university admission through better grades but also to prepare for success during tertiary education and beyond

■ enhance further the coherence and relevance of your IB Diploma Programme experience.

The five approaches to learning develop the following skills:

■ thinking skills

■ social skills

■ communication skills

■ self-management skills

■ research skills.

Practical activities clearly allow you to interact directly with natural phenomena, explore a topic and examine specific research questions. All practical skills covered in this book can be viewed in the context of ATLs. They also give you the opportunity to develop and use IB terminology. These practical skills are:

■ research skills to help you find out appropriate methods to investigate specific research questions, and put your investigation in the context of the wider scientific community

■ thinking skills to help you design investigations, collect and analyze data, and then evaluate your results

■ social skills to help you collaborate with your peers

■ communication skills to help you present your findings effectively and concisely

■ self-management skills to help you plan your time and meet deadlines.

The IB learner profile

The IB Chemistry course is closely linked to the IB learner profile (Table 6). By following the course, you will have engaged with all attributes of the IB learner profile: the requirements of the internal assessment provide opportunities for you to develop every aspect of the learner profile.

Learner profile attribute	Relevance to IB chemistry syllabus
Inquirers	Practical work and internal assessment
Knowledgeable	Links to international-mindedness
	Practical work and internal assessment
Thinkers	Links to theory of knowledge
	Practical work and internal assessment
Communicators	External assessment (examinations)
	Practical work and internal assessment
Principled	Practical work and internal assessment
	Ethical behaviour
	Academic honesty
Open-minded	Links to international-mindedness
	Practical work and internal assessment
	The group 4 project
Caring	Practical work and internal assessment
	The group 4 project
	Ethical behaviour
Risk-takers	Practical work and internal assessment
	The group 4 project
Balanced	Practical work and internal assessment
	The group 4 project
Reflective	Practical work and internal assessment
	The group 4 project

Table 6 Relevance of the IB learner profile to the IB Chemistry syllabus

The internal assessment

The internal assessment forms 20 % of your final mark, with the external examinations (Papers 1, 2 and 3) forming the remaining 80 % of your mark. The assessment and the assessment criteria are the same for both standard level and higher level chemistry.

Criterion	Personal engagement	Exploration	Analysis	Evaluation	Communication	Total marks available
Marks available	2	6	6	6	4	24

Table 7 Marking criteria for the chemistry internal assessment

Your internal assessment mark is based upon one scientific investigation known as the individual investigation. This will involve 10 hours of work and generating a word-processed report of up to 12 pages.

This will be marked out of a maximum of 24 marks based upon the five group 4 assessment criteria (Table 7). This will then be scaled to a mark out of 20. Your individual investigation will be marked internally by your IB Chemistry teacher but moderated externally (re-marked) by an experienced IB Chemistry teacher appointed by the IBO.

There are separate chapters for each of the internal assessment criteria (Chapters 6–10). Checklists at the end of each criterion chapter will help you to ensure that your internal assessment report matches the requirements of the group 4 assessment criteria.

Planning an internal assessment

There are no IB requirements in terms of planning, a time line or documentation, but you have to complete a preliminary internal assessment proposal. This may require you to suggest a research question and **methodology**, carry out a **risk assessment** and complete a requisition for apparatus, instruments and materials for preliminary work.

Setting up a schedule

It may also be helpful to set up a time line with start dates and deadlines for each part of your individual investigation, if your school has not done this. A sample timeline is shown in Table 8.

Stage in investigation	Start date	Task	Deadline date
Planning 1		Read Chapters 6 and 7 in this guide.	
Planning 2		Decide on the research question and carry out background research, identify and classify all the variables, formulate a hypothesis (if appropriate), outline your methodology and data collection and processing.	
Planning 3		Prepare a risk assessment for these experiments and show your chemistry teacher the completed risk assessment form.	
Planning 4		Ensure that the apparatus, instruments and chemical reagents you need will be available in your school chemistry laboratory.	
Practical		Complete the experimental work safely and collect the raw data in the time allocated. Allow sufficient time for preliminary work, and to carry out duplicates, extending the range of data collected. Document any alterations to your plan as soon as they occur; if necessary make alterations to the supporting theory.	
Report 1		Hand in the first draft and consult with your chemistry teacher.	
Report 2		Submit the final draft after an online plagiarism check.	

Table 8 An example internal assessment timeline

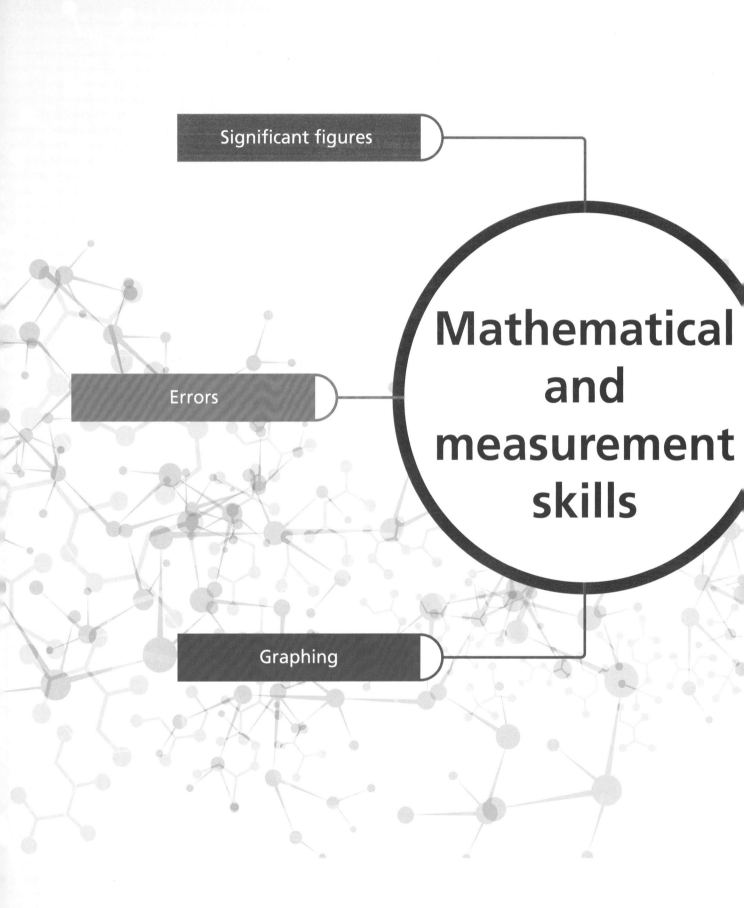

Experimental skills

Significant figures

Errors

Mathematical and measurement skills

Graphing

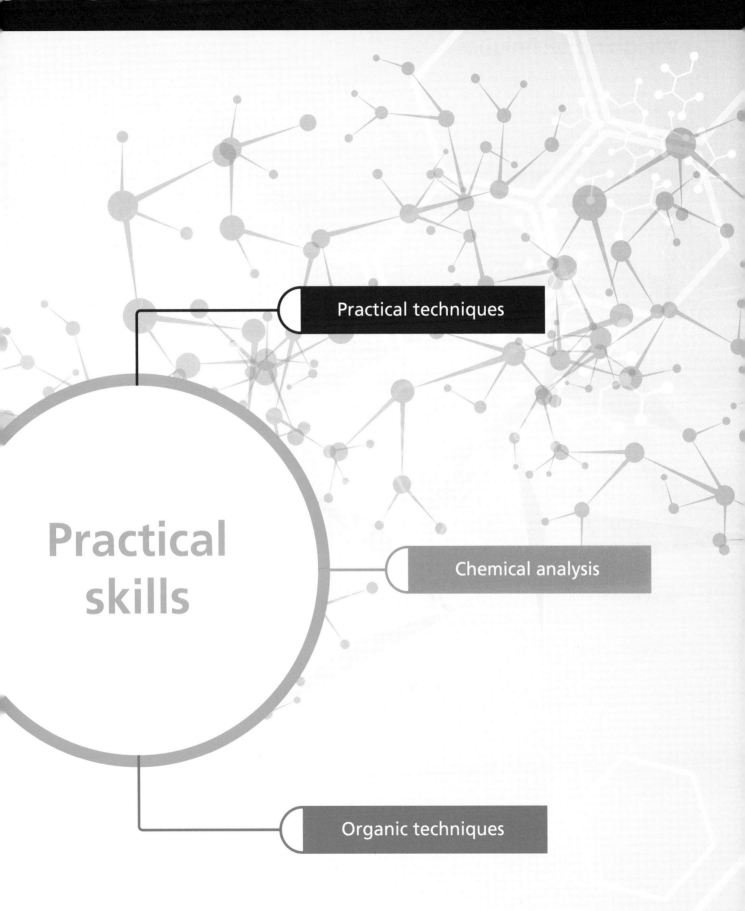

and abilities

Practical techniques

Practical skills

Chemical analysis

Organic techniques

1 Practical techniques

Weighing techniques

- Weigh by difference if you need to transfer the substance into another weighing vessel, or use the tare function of the balance to weigh the substance directly into a weighing bottle.

- If working with an analytical balance, close the doors and let the reading stabilize before recording it.

- Allow hot samples to cool before weighing.

Expert tip

The instructions to weigh accurately about 0.2 g of a specified chemical with an analytical electronic balance (with 4 dp) can be misinterpreted. The important terms are *weigh accurately* and *about 0.2 g*, which specifies the precision (0.0001 g) and a broad range. Any accurate mass within 10 % or ±0.02 g of 0.2 g is acceptable.

Tare the weighing bottle, zero the balance, add a small amount of the solid and record its mass. Taring involves subtraction of the mass of the vessel from the mass of the sample and weighing bottle to determine the mass of the sample.

Reset the analytical balance so the digital readout is 0.0000 g, then add the solid to the weighing bottle until the total mass is in the range, 0.20 ± 0.02 g (or 0.18 g to 0.22 g). You should record the sample mass to 0.0001 g, for example, 0.1968 g. This is an accurately known sample of about 0.2 g.

Collecting gases

The method of gas collection (Table 1.1) depends on two properties of the gas:

- its solubility in water (which varies with temperature)

- its density compared to air (Table 1.2).

■ ACTIVITY

1 Explain how you can know whether a gas is heavier or lighter than air. Explain why downward displacement of water is not an ideal method for collecting carbon dioxide gas.

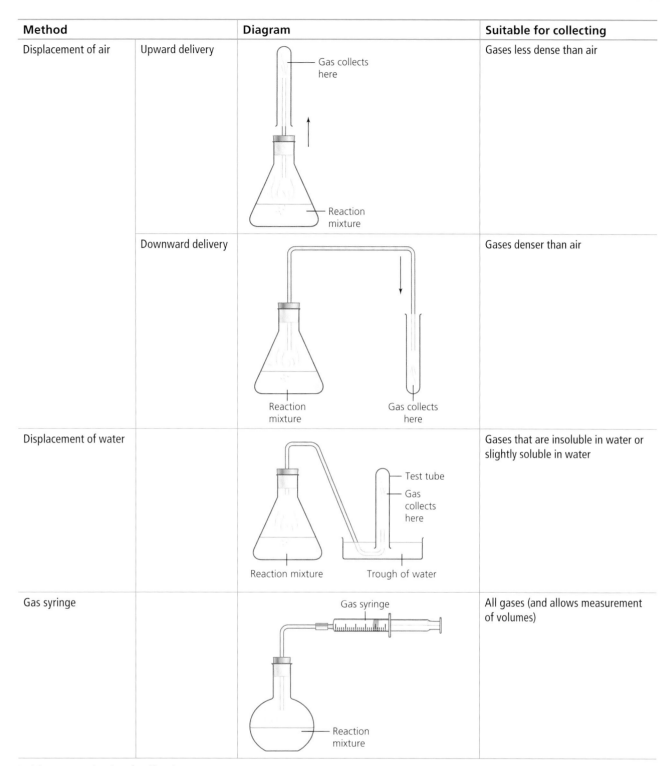

Method		Diagram	Suitable for collecting
Displacement of air	Upward delivery		Gases less dense than air
	Downward delivery		Gases denser than air
Displacement of water			Gases that are insoluble in water or slightly soluble in water
Gas syringe			All gases (and allows measurement of volumes)

Table 1.1 Methods of collecting gases

	Soluble in water	Slightly soluble in water	Insoluble in water
Less dense than air	Ammonia, NH_3	–	Hydrogen, H_2
Similar in density to air	–	Nitrogen, N_2 Carbon monoxide, CO	–
Denser than air	Chlorine, Cl_2 Hydrogen chloride, HCl Nitrogen dioxide, NO_2 Sulfur trioxide, SO_3	Argon, Ar Carbon dioxide, CO_2 Oxygen, O_2	–

Table 1.2 Properties of selected gases

Drying gases

A U-tube (Figure 1.1) with a water-absorbing solid chemical is used to dry a gas (Table 1.3). It is inserted between the reaction container and the gas collection system. Gloves and safety glasses must be worn at all times when packing or emptying a drying tube.

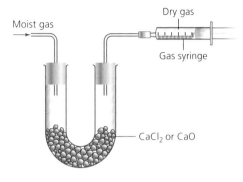

Figure 1.1 U-tube with drying agent

Drying agent	Suitability	Important notes
Anhydrous calcium chloride	Most gases except ammonia	Use lumps or granules of the drying agent
Calcium oxide	Suitable for basic oxides and neutral gases	
	Not suitable for acidic gases, such as HCl	

Table 1.3 Drying gases

Expert tip

To minimize experimental errors or increase reliability:

- you should clamp the apparatus properly using retort stands
- you must ensure the apparatus is airtight to minimize the risk of gas escaping, so that the results you obtain are accurate
- when you swirl the conical flask to mix the reactants, you should be aware that the swirling could affect the volume of gas that is collected and measured
- since gases are compressible, heat released during the reaction will affect the volume of the gas collected. It is therefore important to let the system reach equilibrium in terms of temperature and pressure before recording the gas volume.

Ideas for investigations involving gas collection

- Analyze mixtures of salts.
- Determine the value of the gas constant.
- Determine the stoichiometry of a decomposition reaction.
- Determine the amount of water of crystallization in a hydrated salt or weak acid.

Gas stoichiometry

Accurate measurements of gas volumes use a eudiometer (Figure 1.2): a graduated glass tube closed at one end.

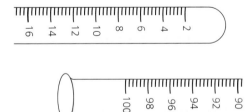

Figure 1.2 The open and closed ends of two eudiometers

A eudiometer can be used to determine the molar mass of a gas, or the molar gas volume (or the gas constant, R, if pure gas of a known molar mass is used) via the ideal gas law. Once the amount of gas has been calculated, the molar mass can be calculated via the mass.

Temperature and pressure must be measured as well and, since the volume of gas is measured over water, there will be two gases in the eudiometer: the gas from a chemical reaction or released from a gas cylinder, and water vapour. The pressure in the eudiometer is the total pressure of the gas/water vapour mixture.

■ **ACTIVITIES**

Butane gas (C_4H_{10}) was released from a butane lighter into a eudiometer filled with water. The butane released contained small quantities of other hydrocarbons. The pressure of the surrounding air was recorded using a barometer. The raw data from the experiment is tabulated below. (1 Pa = 0.007 500 62 mm Hg) ideal gas equation: pV = nRT, where T is expressed in kelvin

Initial mass of the butane lighter	16.423 g ± 0.001 g
Final mass of the butane lighter	16.344 g ± 0.001 g
Temperature of the water (surrounding the eudiometer)	25.0 °C ± 0.5 °C
Volume of butane in the eudiometer	30.0 cm³ ± 0.1 cm³
Atmospheric pressure	766.10 mm Hg ± 0.01 mm Hg

2 Determine the mass (in grams) of butane released from the butane lighter.

3 Determine the pressure (in Pascal) of dry butane present in the eudiometer. Apply Dalton's law of partial pressure (p atmosphere = p butane + p water vapour). The vapour pressure of water at 21 °C is 23.76 mm Hg.

4 Determine the amount (in moles) of butane gas present.

5 Determine the experimental molar mass of butane.

6 Determine the **percentage error**. Use the IB Data booklet to calculate the molar mass of butane.

7 State two assumptions of this experimental approach (apart from ideal gas behaviour).

8 Identify at least one potential source of error in the experimental procedure that may lead to a loss of accuracy.

Electrochemical cells

You will need to take some precautions to obtain accurate values of cell potentials. You must clean the metal electrodes thoroughly. You should rub them with fine emery paper to remove any oxidation products, and then rinse them with distilled water and dry them in an oven. To make them grease-free, rub them with cotton wool soaked in propanone (acetone) before rinsing. Handle them by the edges.

If your meter reading is below zero, reverse the connections to your electrochemical cell (Figure 1.3). Record the voltage and note which of the electrodes is positive and which is negative. A salt bridge can be made from filter paper soaked in saturated potassium nitrate, but better results will be obtained using a U-tube with the electrolyte in agar gel.

> ### Examiner guidance
>
> Reactive metals such as magnesium will undergo a slow direct reaction with water to form hydrogen, reducing the voltage of the electrochemical cell to below its theoretical value (calculated from standard electrode potentials).

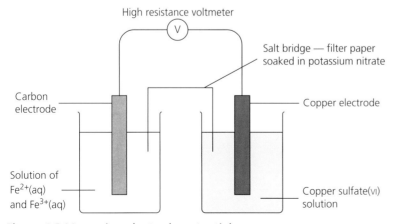

Figure 1.3 Measuring electrode potentials

■ ACTIVITIES

Consider a simple Daniell cell that consists of zinc immersed in zinc ions and copper in copper(II) ions connected by a salt bridge composed of potassium nitrate solution in a U-tube filled with agar. Assume standard thermodynamic conditions.

9 List all the variables that **could** affect the voltage generated by a Daniell cell.

10 Consult a text book or the internet to find the Nernst equation and state two variables that **will** change the cell voltage of the Daniell cell.

11 Use the Nernst equation to determine the cell voltage for a copper/magnesium cell at different concentrations of the two ions at 25 °C.

$[Mg^{2+}(aq)]/mol\ dm^{-3}$	1.000	0.100	0.010	0.001	1.000	1.000	1.000
$[Cu^{2+}(aq)]/mol\ dm^{-3}$	1.000	1.000	1.000	1.000	0.100	0.01	0.001
Cell voltage/V							

Examiner guidance

Strictly speaking, the cell voltage is the cell electromotive force (emf). It is not a force but the potential difference ('voltage') when no current flows. It is measured using a very high resistance voltmeter. The voltage across the cell is less than the emf when the battery supplies current.

Electrolytic cells

The essential issues in an electrolytic investigation are: the quantity (mass, volume or amount) of substance that can be obtained from the anode, cathode or both; the amount of charge that has passed through the circuit ($Q = I \times t$); and how to connect the amount of substance that is 'collected' at the electrode with the quantity of charge that has passed through the circuit.

Ideas for investigations involving electrochemistry

- Determine the effect of concentration or temperature on cell potential of a simple cell or model lead–acid battery.
- Investigate electroplating (including anodizing).
- Investigate electrolysis products from sodium chloride solution of different concentrations.
- Investigate oxygen evolution at graphite electrodes.
- Determine the charges of ions and the Avogadro constant.

■ ACTIVITIES

A student conducted an electroplating experiment with copper(II) sulfate solution with copper electrodes to determine the magnitude of the Faraday constant, F, and collected the following raw data. The half equation at the cathode is: $Cu^{2+}(aq) + 2e^- \rightarrow Cu(s)$.

Current passed: 0.30 A ± 0.02 A

Duration of experiment: 300.0 s ± 0.2 s

Mass of copper cathode:

        at start of the experiment: 2.320 g ± 0.001 g

        at end of the experiment: 2.350 g ± 0.001 g

12 Calculate the amount of copper atoms (in moles) deposited on the cathode surface.

13 Calculate the value of the Faraday constant.

14 Calculate the absolute random uncertainty for the experimentally determined value of the Faraday constant.

15 The literature value for the Faraday constant is 96 485 C mol⁻¹. Calculate the percentage error of the Faraday's constant obtained experimentally.

Use of pH probe and meter

Take care to clean the electrode thoroughly by rinsing it with distilled water from a wash bottle every time you change solution. Do **not** leave the electrode unimmersed for longer than you need or let the electrode dry out: salt deposits will form and interfere with the electrode response.

Standardize the meter (manually or the meter may do it automatically) at a particular pH and temperature using the buffer(s) provided. Keep stirring the sample during **calibration** but make sure that the end of the probe stays wet.

Do **not** touch the electrode or move about near to it while a reading is being taken. If you are using a magnetic stirrer, switch it **off** when you take the readings. The electrode needs time to reach equilibrium in solution, so do **not** rush to record the measurement, but wait until the meter reading becomes steady.

Expert tip

The buffer provided should reflect the nature of the solutions whose pH values you are going to measure. For example, if you are investigating basic solutions then a basic buffer solution (for example, pH 9.0) should be provided. However, calibrating with two buffers allows the 'slope' of the pH meter to be set so that it reads correctly at all pH values.

Indicators

Acid–base indicators (usually weak acids) change colour in a pH-dependent way. They may be added in small amounts to an aqueous solution, or they can be used in paper-strip form. Each indicator dye usually changes colour over a restricted pH range (Table 1.4).

Indicator	pK_a	pH range	Colour change Acid	Colour change Alkali
Methyl orange	3.7	3.1–4.4	Red	Yellow
Bromophenol blue	4.2	3.0–4.6	Yellow	Blue
Bromocresol green	4.7	3.8–5.4	Yellow	Blue
Methyl red	5.1	4.4–6.2	Red	Yellow
Bromothymol blue	7.0	6.0–7.6	Yellow	Blue
Phenol red	7.9	6.8–8.4	Yellow	Red
Phenolphthalein	9.6	8.3–10.0	Colourless	Pink

Table 1.4 Acid–base indicators

Universal indicator is a mixture of dyes that changes colour a number of times between high and low pH values (Figure 1.4).

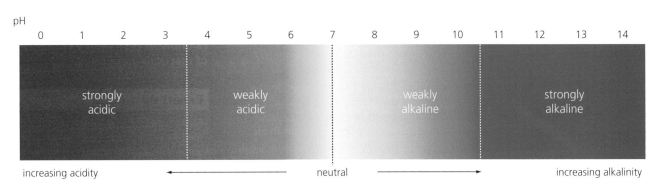

Figure 1.4 The pH scale and the colours of universal indicator solution

Examiner guidance

Acid–base indicators are not suitable for accurate pH measurement as they are affected by oxidizing and reducing agents and salts. They can estimate the approximate pH of a solution and determine a change in pH (for example, at the end-point of a titration). The pH range over which the indicator changes colour must fall within the (almost) vertical region of the titration curve for it to be suitable.

Buffers

When creating a buffer solution, you need to consider the ratio of acid and conjugate base required to give the correct pH and the amount of buffering needed, which depends on the amounts of acid and base. To make a buffer, add most of the volume of double distilled water to a beaker, then add the weighed chemicals. When the chemicals have been mixed and dissolved, check and adjust the pH. Pour the buffer into a volumetric flask, then add water to make up the solution to the final volume needed. Shake and place the buffer in a clean container, then seal and label with name of buffer, date prepared and your name.

■ ACTIVITIES

16 Describe the preparation of 1000 cm³ of a 0.2 mol dm⁻³ ethanoate buffer of pH 5. The pK_a of ethanoic acid is 4.76.

Ideas for investigations involving buffers
• Prepare buffer solutions and investigate their ability to withstand changes in pH. • Use a buffer solution to determine the concentration of a weak acid or base. • Investigate the effect of temperature on the pH of a buffer or its buffering capacity.

Chemical energetics (calorimetry)

▍ Enthalpies in solution

Determinations of enthalpies of neutralization, replacement, precipitation and neutralization are normally carried out in a thermally insulated plastic cup to minimize heat loss to the surroundings (Figure 1.5). They involve reacting together volumes of known concentration and recording temperatures against time or the maximum or minimum temperature reached.

■ ACTIVITIES

Ammonium chloride is commonly used in instant cold packs. The cold pack contains water, and in the water is another pouch containing the ammonium salt. When the pack is squeezed, the inner pouch is broken, releasing the salt, which quickly dissolves and lowers the temperature of the pack.

A data book gives the following information:

• solubility of ammonium chloride at 25 °C = 6.95 mol dm⁻³
• 4.18 J are required to raise the temperature of 1.0 cm³ of the solution by 1 °C
• the enthalpy change of solution of ammonium chloride is approximately +15 kJ mol⁻¹.

17 Calculate the maximum mass of ammonium chloride that can be added to 100 cm³ of water in the simple calorimeter shown in Figure 1.5. Suggest a minimum mass of ammonium chloride that could be used. **Justify** your choices with relevant calculations, stating any assumptions you have made.

18 Draw a sketch of the graph that you expect to obtain in the experiment. Indicate clearly on the graph the initial and final temperatures that you would read.

Examiner guidance

Recording temperatures at regular time **intervals** helps to correct for heat loss (exothermic reactions) and heat gain (endothermic reactions). When an **extrapolation** (Figure 1.6) is done, the difference between the initial and the maximum temperature is the theoretical change of temperature without any heat loss to the surroundings.

Expert tip

Citric and phosphate buffers form insoluble complexes with dipositive cations, and phosphate ions can also act as a substrate activator or inhibitor of certain enzymes. Zwitterionic buffers, for example, HEPES, are preferred for biochemical investigations.

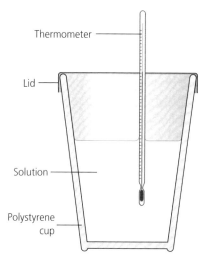

Figure 1.5 A simple polystyrene calorimeter (to measure enthalpies of reactions in solution)

Expert tip

Make sure that you stir the reaction mixture gently with the thermometer while the reaction is occurring and avoid loss of liquids through splashing.

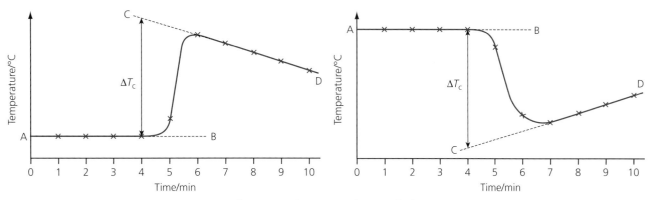

Figure 1.6 Temperature correction curves for an exothermic and an endothermic reaction

To minimize experimental errors or increase reliability when carrying out an enthalpies in solution experiment, bear in mind the following points.

- Some compounds, such as group 1 hydroxides, are hygroscopic and absorb water vapour from moist air. Weigh the compound quickly or cover after weighing to avoid absorption of water vapour from the air.
- When making solutions, if the compounds that dissolve release or absorb heat energy, wait for the solution to reach thermal equilibrium with the surroundings before using it.
- Because heat is easily lost to the surroundings through convection, carry out a thermometric experiment in a draught-free laboratory.
- Cover the polystyrene cup with a lid to provide thermal lagging to minimize heat loss to the surroundings.
- Measure the initial temperatures of the two solutions before mixing. Calculate the average temperature if necessary.
- Use a burette or pipette to measure volumes of solution. These instruments are more accurate than measuring cylinders.
- If a more accurate measurement of the temperature is needed, use a data-logger with a temperature sensor that has a high sensitivity and high precision.

Enthalpies of combustion

Enthalpies of combustion of volatile organic liquids can be determined using a spirit burner placed under a copper calorimeter (Figure 1.7). A known mass of pure liquid is combusted and the temperature rise of the water in the calorimeter recorded. Not all of the heat energy that is released from the combustion process enters the water. Some of the heat energy is absorbed by the copper calorimeter while some of the heat energy is lost due to convection in the air.

> **Key definition**
>
> **Enthalpy change of combustion** – the heat energy released when one mole of pure compound is combusted in excess oxygen under standard conditions.

Examiner guidance

The copper calorimeter absorbs some heat but the amount of heat absorbed by the calorimeter can be calculated and the accuracy of the experiment improved. In a separate experiment, a fuel with a known enthalpy of combustion can be burnt and this can be used to calculate the heat capacity of the copper calorimeter.

■ ACTIVITIES

19 Explain why the copper can should not be replaced by a glass beaker.

20 Explain why the thermometer should not touch the base of the copper calorimeter.

■ ACTIVITIES

Cyclohexa-1,3-diene (molar mass 80 g mol^{-1}) is a highly flammable cycloalkene that is a colourless clear liquid under standard conditions. It is a potential fuel that can be used by burning it in air.

21 Define the term **standard enthalpy change of combustion** and write a balanced equation for the complete combustion of cyclohexa-1,3-diene.

22 Using the apparatus shown in Figure 1.7, give a brief outline of the procedure you would follow in order to determine the standard enthalpy change of combustion of cyclohexa-1,3-diene. Make clear references to the amount of each substance you would use and the measurements you would record during the experiment.

23 Outline how you would use your experimental results to determine the enthalpy change of combustion of cyclohexa-1,3-diene. You may assume that the specific heat capacity of water is 4.18 J g^{-1} K^{-1}.

24 Other than heat loss to the surroundings, state another assumption you have made in your calculations and discuss how it will affect the standard enthalpy change of combustion calculated from the experimental data.

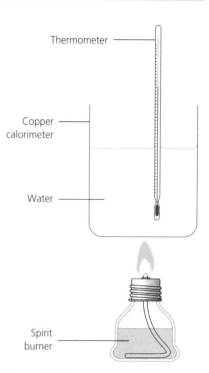

Figure 1.7 Simple apparatus used to measure enthalpy changes of combustion of pure volatile liquids

Expert tip

To minimize experimental errors or increase reliability when carrying out an enthalpy of combustion experiment:

● the metal calorimeter should be lagged at the sides, not at the bottom, which needs to be in contact with the flame

● the flame cannot be too far away from the metal calorimeter as this will lead to more heat loss by convection

● the accuracy of the measured mass of the spirit lamp is limited by the burning of the wick. This cannot be avoided.

Ideas for investigations involving enthalpy

● Determine the enthalpy change of solutions of salts.

● Determine enthalpy changes via Hess's law (for example, enthalpy changes of hydration for salts and enthalpies of formation for metal oxides).

● Determine the enthalpy change of displacement for a metal reacting with cations in aqueous solution.

● Determine the enthalpy of combustion of a range of related organic compounds.

● Determine the enthalpy change between bases and acids.

Chemical kinetics

■ Reaction rates

Chemical kinetics is the study of reaction rates. Reaction rates provide data about how fast a chemical process occurs, as well as suggesting the mechanism.

The rate of reaction is a measure of the change in concentration of reactants or products over time. The rate measured at the beginning of a reaction is known as the **initial rate**; the rate measured at any point in time while the chemical reaction is in progress is known as the **instantaneous rate**; the rate measured over an interval of time is called the **average rate**.

By determining the dependence of the initial rate on the initial concentrations of reactants, we can determine the experimental rate law for the reaction.

The instantaneous rate is calculated from the slope of a tangent drawn at any point on the graph of concentration versus time. The slope of the tangent taken at the initial point of the graph is assumed to be equal to the initial rate (Figures 1.8 and 1.9).

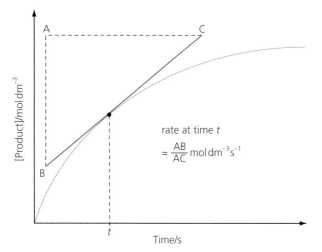
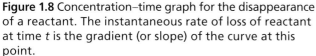

Figure 1.8 Concentration–time graph for the disappearance of a reactant. The instantaneous rate of loss of reactant at time t is the gradient (or slope) of the curve at this point.

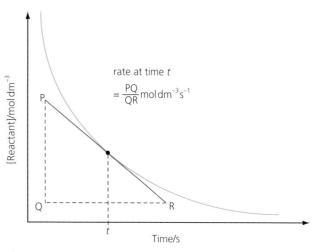

Figure 1.9 Concentration–time graph for the formation of a product. The instantaneous rate of formation of product at time t is the gradient (or slope) of the curve at this point.

Examiner guidance

The average rate is the mean of several instantaneous rates taken over a period of time. Note that the instantaneous rate of a reaction decreases as time progresses because the concentrations of the reactants decrease. This means that the average rate taken over a longer period will have a smaller value compared to one taken over a shorter period after the reaction begins.

Investigating rates of reaction

Investigating the rate of a reaction usually involves a method that indirectly measures the change in concentration of products or reactants with changes in time. These methods include:

■ measuring the total volume of gas collected at various time intervals using a gas syringe connected to the reaction vessel

■ measuring the total mass of a reaction mixture that releases a gas using an electronic balance to measure mass at various time intervals

■ the clock method: observing precipitation/colour change and noting the specific time at which the precipitation or colour change occurred

■ for reactants or products that are coloured, measuring the intensity of the colour using a colorimeter at various time intervals

■ recording the electrical conductivity of an aqueous solution with a conductivity probe at various time intervals

■ recording the optical activity with a polarimeter at various time intervals

■ recording gas pressure at various time intervals

■ the titration method: removal of reaction samples at timed intervals to determine the amount of a specific reactant or product via an acid–base or redox titration.

Examiner guidance

In the titration method, the reaction sample is 'quenched' to slow or stop the reaction. This can be done by dilution with iced water or by chemically removing one of the reactants (for example, by adding hydrogencarbonate ions to neutralize protons in the reaction mixture).

■ **ACTIVITIES**

25 Find out about the use of a dilatometer.

26 Suggest the advantages and disadvantages of following a physical property.

These methods require the collection of several data points during one run of the experiments, at various suitable time intervals. These data points are used (or processed) to generate a graph of concentration versus time, which can be used to determine the order and initial rate.

> ### Examiner guidance
>
> For the 'clock' method, a number of experiments are carried out; usually, the concentration of a specific reactant is varied while the concentrations of the other reactants are kept constant. The time is recorded until a rapid colour change is observed. The reciprocal of the time (average rate) is approximately proportional to the initial rate. Measurements that are precise and obtained by the same student with the same apparatus are described as **repeatable**.

To determine the rate equation, there are two approaches: single reaction mixture (for example, the sampling method) and multiple reaction mixtures (for example, the clock reaction/initial rate methods).

■ The sampling method

The sampling method uses a single reaction mixture such as those listed in Table 1.5. You sample equal portions of a total volume of the solution at regular time intervals from the reaction mixture. You then analyze the sampled solutions to determine how their properties change over time. You must quench the reaction in the samples if chemical analysis is to be carried out.

	Can be titrated with	**Examples**
Acids	Bases	NaOH, Na_2CO_3, $NaHCO_3$
Bases	Acids	HCl
Reducing agents	Oxidizing agents	H_2O_2, $C_2O_4^{2-}$ with MnO_4^-
Oxidizing agents	Reducing agents	I_2 with $S_2O_3^{2-}$

Table 1.5 Examples of single reaction mixtures

■ Initial rate method

The initial rate method uses multiple reaction mixtures. You should measure the time taken for a fixed amount of reactant to be used or a fixed amount of product to be formed or for the reaction to go to completion (to reach an identifiable point in the reaction). You can monitor this by recording the time taken for a fixed amount of chemical (for example, iodine or sulfur) to hide a cross placed under the reaction flask, or by the addition of a suitable reagent to cause a sharp colour change (for example, record the time taken for iodine to react with a fixed amount of thiosulfate ions in the presence of starch).

- To determine the value of the individual order with respect to A: carry out two sets of experiments in which the concentration of B is the same and the concentration of A changes. Measure how the rate changes with the concentration of A being changed.

- To determine the value of the individual order with respect to B: carry out two sets of experiments in which the concentration of A is the same and the concentration of B changes. Measure how the rate changes with the concentration of B being changed.

■ **ACTIVITY**

27 Outline the relative advantages and disadvantages of clock/initial rate experiments.

■ ACTIVITIES

Consider the following kinetic data for the reaction:

$$S_2O_8^{2-}(aq) + 3I^-(aq) \rightarrow 2SO_4^{2-}(aq) + I_3^-(aq)$$

Experiment	$[S_2O_8^{2-}]/mol\,dm^{-3}$	$[I^-]/mol\,dm^{-3}$	Initial rate/$mol\,dm^{-3}\,s^{-1}$
1	0.038	0.060	1.5×10^{-5}
2	0.076	0.060	2.8×10^{-5}
3	0.038	0.120	2.9×10^{-5}

28 Calculate the order with respect to both reactants and write the rate equation for this reaction.

29 Calculate the rate constant and the reaction rate when the concentration of both reactants is $0.050\,mol\,dm^{-3}$. Use the results of the first experiment.

■ Continuous (bulk) method

- To determine the value of the order with respect to A: perform one set of experiments in which the concentration of B is much higher than the concentration of A. Measure the concentration of A with respect to time.

- To determine the value of the order with respect to B: perform one set of experiments in which the concentration of A is much higher than the concentration of B. Measure the concentration of B with respect to time.

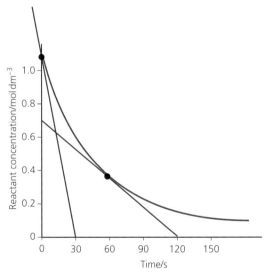

Figure 1.10 Drawing tangents at different times to find the instantaneous rates

However, drawing tangents to a curve of this kind is difficult, and it is easier and more accurate to use integration of the rate equation to test likely reaction orders. A graphical method (Table 1.6) can be used for determining zero-order, first-order and second-order rates from the continuous (bulk) method.

Overall reaction order	Zero order A → products	First order A → products	Simple second order* A → products A + B → products
Rate expression	Rate = $k[A]_0$**	Rate = $k[A]$	Rate = $k[A]^2$ Rate = $k[A][B]$
Data to plot for a straight-line graph	[A] versus t	ln [A] versus t	$\dfrac{1}{[A]}$ versus t
Slope or gradient equals	$-k$	$-k$	$+k$
Changes in the half-life as the reactant is consumed	$\dfrac{[A]_0}{2k}$ $t_{1/2}$ becomes shorter	$\dfrac{\ln 2}{k}$ $t_{1/2}$ is constant	$\dfrac{1}{k[A]_0}$ $t_{1/2}$ becomes longer
Units of k	$mol\,dm^{-3}\,s^{-1}$	s^{-1}	$dm^3\,mol^{-1}\,s^{-1}$

* A simple second-order reaction is a reaction which is second order with respect to one reactant; that is, rate = $k[A]^2$.
** $[A]_0$ is the initial concentration (at time $t = 0$).

Table 1.6 Summary of graphical methods to find orders and half-lives

■ **ACTIVITIES**

To investigate the effect of concentration changes on the rate of a reaction, an experiment was carried out using magnesium ribbon of mass 0.125 g placed in a conical flask with 15.00 cm³ of 0.5000 mol dm⁻³ hydrochloric acid. The progress of this reaction was followed by measuring the gas produced with a gas syringe.

30 Show that the hydrochloric acid is the limiting reagent and calculate the maximum volume of hydrogen gas (cm³) produced (at STP).

31 Suggest why the magnesium should be cleaned with fine sandpaper before the reaction.

32 The chemical literature states the reaction is first order with respect to hydrochloric acid. Outline how you would use the raw data to support this hypothesis.

33 State one safety risk with hydrogen gas.

■ **ACTIVITIES**

Hydrochloric acid reacts with sodium thiosulfate to produce sulfur, sulfur dioxide and water. The order of the reaction can be determined using the initial rate method. This involves judging relative rates by the reciprocal of the time taken to obscure a cross on a piece of paper placed under a conical flask. You have 0.400 mol dm⁻³ sodium thiosulfate, 2.00 mol dm⁻³ hydrochloric acid and access to laboratory equipment and apparatus.

$$2H^+(aq) + S_2O_3^{2-}(aq) \rightarrow S(s) + SO_2(g) + H_2O(l)$$

34 Complete the table below:

Volume of 0.400 mol dm⁻³ sodium thiosulfate/cm³ ± 0.05 cm³	Volume of 2.00 mol dm⁻³ hydrochloric acid/cm³ ± 0.05 cm³	Volume of water/cm³ ± 0.05 cm³	Time/s
40.00	30.00	0.00	–
30.00	20.00	20.00	–
40.00	15.00		–
40.00	5.00		–
30.00	10.00		–

35 Explain why the volume of water is varied in each experiment.

36 Suggest the expected relationship between the volume of sodium thiosulfate and its concentration.

37 Suggest the expected relationship between the rate of reaction and the time taken for the cross to be hidden by the precipitate of sulfur.

38 A gas with a foul rotten egg smell is produced in a side reaction. Suggest the name of this gas and how you can test for it.

Ideas for investigations involving kinetics

- Determine the rate of reaction and order between propanone, hydrogen ions and iodine.
- Determine the rate of reaction between potassium manganate(VII) and ethanedioate ions.
- Determine the rate of reaction between potassium bromate(V) with potassium bromide in acidic conditions and in the presence of phenol.
- Determine the rate of reaction between potassium iodide and hydrogen peroxide in acidic conditions.
- Determine the rate of hydrolysis of benzenediazonium chloride.
- Investigate an iodine or bromine clock reaction.
- Investigate light emission from luminol.
- Investigate the bleaching of dyes (via colorimetry).
- Investigate the reaction between 2,3-dihydroxybutanedioate (potassium sodium tartrate) and hydrogen peroxide in the presence of cobalt(II) ions.
- Determine the activation energy or order of reaction for the reduction of iodide ions by peroxydisulfate ions (via a clock reaction).
- Investigate the effect of zeolite HY as a catalyst for the synthesis of an ester.

Chemical equilibria

To determine the equilibrium constant, the stoichiometry (balanced equation) of the chemical reaction must be known. You must have an analytical method for measuring the concentrations of reactants and products at equilibrium.

If the initial concentrations of all species are known, it may be possible to determine the value of the equilibrium constant by measuring the equilibrium concentration of only one of the species involved in the reaction (since the final concentrations can be determined using stoichiometry calculations, via the ICE method).

The general procedure is that the concentration is measured for a series of solutions with known concentrations of the reactants. Often a titration is carried out with one or more reactants in the flask and one or more reactants in the burette. Knowing the concentrations of reactants initially in the flask and in the burette, all concentrations can be derived as a function of the volume (or mass) of **titrant** added.

> **Key definition**
> **Titrant** – the substance added during the titration.

For example, mixtures containing various amounts of water, ethyl ethanoate, ethanoic acid and ethanol, together with concentrated hydrochloric acid (protons act as a catalyst), can be prepared. Equilibrium is approached from various starting points and the mixtures are then titrated with alkali. By subtracting the amount of alkali needed to neutralize the catalyst, the amount of ethanoic acid present at equilibrium can be deduced. The amounts of the other reactants can then be determined.

Examiner guidance

In this example involving esterification, the equilibrium constant, K_c, has no units, but it is good practice to convert the values into concentrations before completing the calculation.

■ ACTIVITIES

A student carried out a series of experiments to determine the equilibrium constant for the esterification reaction between ethanol and ethanoic acid to form the ester ethyl ethanoate and water:

$$C_2H_5OH(l) + CH_3COOH(l) \rightleftharpoons CH_3COOC_2H_5(aq) + H_2O(l)$$

Four conical flasks containing different proportions of ethanol, glacial (pure) ethanoic acid and $1.00 \, mol \, dm^{-3}$ hydrochloric acid were set up and left stoppered for two weeks to reach equilibrium. They were shaken four times a day. The contents of each flask were then titrated with a $1.00 \, mol \, dm^{-3}$ solution of potassium hydroxide (freshly prepared) and the results recorded.

	Flask 1	Flask 2	Flask 3	Flask 4
Mass of ethanol/g ± 0.001 g	3.000	3.000	3.000	4.000
Mass of glacial ethanoic acid/g ± 0.001 g	3.000	3.000	3.000	2.500
Volume of hydrochloric acid/cm³ ± 0.10 cm³	6.50	6.50	6.50	4.50
Potassium hydroxide solution titre/cm³ ± 0.10 cm³	33.10	37.10	33.20	18.50

39 Suggest what indicator could be used in the titrations.

40 Explain why the flasks were stoppered and shaken four times a day.

41 If the student had measured the volume of the ethanol instead of weighing it, state what information would be needed to convert the volume to a mass.

42 Suggest which data point appears to be anomalous.

43 $K_c = \dfrac{\text{[ester][water]}}{\text{[alcohol][acid]}}$

The student found that the calculated values were all smaller than the literature value in a data book. Suggest why the experimental values for K_c were smaller.

Ideas for possible investigations involving quantitative equilibrium

- Investigate the equilibrium between bismuth(III) oxychloride and bismuth; nitrogen dioxide and dinitrogen tetroxide; iodine monochloride and iodine trichloride; chromate(VI) and dichromate(VI) ions; iron(III) and cyanate ions using colorimetry.
- Determine the solubility product of a sparingly soluble ionic compound.
- Investigate the common ion effect.
- Determine the partition coefficient of ammonia solution with an organic solvent via a titration.
- Investigate the equilibrium between silver ions and iron(II) ions and silver and iron(III) ions using potassium thiocyanate ions to titrate silver ions.

2 Chemical analysis

Volumetric analysis

▨ Standard solutions

- ▪ If there are impurities in the primary standard then the true mass present will be less than the measured mass. The solution will have a concentration less than the calculated value (a **systematic error** (Chapter 4)).

- ▪ The **standard solution** must be stable in air and in aqueous solution, or it will react with oxygen or carbon dioxide in the air or with water (hydrolysis).

- ▪ The standard solution must be soluble in water so that solutions of high concentrations can be prepared.

- ▪ The standard solution should have a large molar mass in order to minimize the percentage random **uncertainty** (Chapter 4) in the mass of substance weighed out.

Some examples of reagents used as primary standards are outlined in Table 2.1.

> **Key definition**
>
> **Standard solution** – a solution with an accurately known concentration, prepared from a primary standard. This is a substance dissolved in a known volume of water to give a standard solution.

Primary standard	Examples
Acid	Hydrated ethanedioc acid, $(COOH)_2.2H_2O$ and potassium hydrogenphthalate: COOK / COOH (benzene ring structure)
Base	Anhydrous sodium carbonate, Na_2CO_3
Oxidizing agent	Potassium dichromate(vi), $K_2Cr_2O_7$; potassium iodate(v), KIO_3
Reducing agent	Sodium ethanedioate, $(COONa)_2$
Complexing agent	Hydrated disodium salt of EDTA: $NaOOCCH_2$ \ CH_2COONa / NCH_2CH_2N $.2H_2O$ / \ $HOOCCH_2$ CH_2COOH

Table 2.1 Common primary standards

▪ ACTIVITY

1 Research the chemical properties of sodium hydroxide and explain why it is not a primary standard.

You need to calculate the mass of the primary standard required from the volume and concentration of solution required. The sample must be dried in a desiccator (Figure 2.1) to remove any water absorbed from the atmosphere. Primary standards can be dried by heating in an oven, but they may decompose at high temperature.

An empty dry weighing bottle is used to weigh out the primary standard, either by difference or by taring (see Chapter 1). The sample of the primary standard is then transferred to a beaker of distilled water and stirred with a glass rod.

Expert tip

The stopper must be removed from the weighing bottle only when necessary to reduce the possibility of it re-absorbing water vapour.

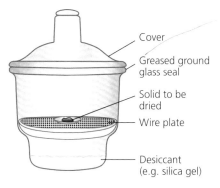

Figure 2.1 A laboratory desiccator

The solution is added to a volumetric flask using a filter funnel (Figure 2.2). The inside of the beaker and stirring rod should be washed with distilled water and the washings transferred. The washing process should be repeated to make sure that all the primary standard has been transferred.

Distilled water is then added to the volumetric flask until the solution is close to the graduation mark. After the funnel is rinsed and removed, distilled water is added using a dropper until the bottom of the meniscus is level with the graduation mark.

Expert tip

The volumetric flask should then be stoppered and turned upside down several times to ensure the solution is thoroughly mixed and is of uniform concentration (homogeneous).

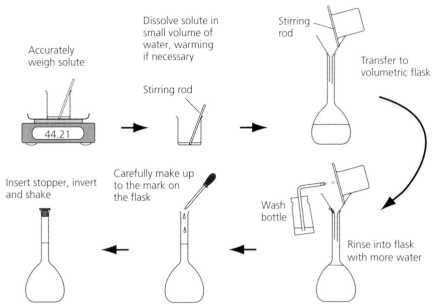

Accurately weigh solute
44.21

Dissolve solute in small volume of water, warming if necessary

Stirring rod

Stirring rod

Transfer to volumetric flask

Wash bottle

Rinse into flask with more water

Carefully make up to the mark on the flask

Insert stopper, invert and shake

Figure 2.2 Preparing a standard solution

Titrations

A clean burette is rinsed with a small volume of the standard solution. It is then tilted to an almost horizontal position and rotated so the solution 'wets' the inside. The tip is rinsed by draining the solution through it. The burette is then clamped vertically and filled with the standard solution slightly beyond the zero mark; the tip is filled by opening the tap.

ACTIVITY

2   Explain the effect on an acid–base titration if the conical flask with the alkali is wet with distilled water before the acid solution is added to it.

With a white piece of paper behind the burette and with your eye level with the top of the standard solution, read the burette from the bottom of the meniscus and record the reading (Figure 2.3).

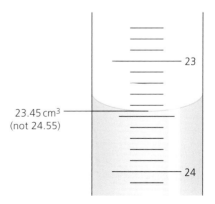

23

$23.45\,cm^3$
(not 24.55)

24

Figure 2.3 Recording a titre volume from a burette

Expert tip

When reading a burette, it is important to first remove the filter funnel you used to fill it. If this is left in place, some drops of solution could drain from it during the titration, leading to a false titre volume.

Use a pipette to transfer a fixed volume, usually $25.00\,cm^3$, of solution of unknown concentration (the analyte) to a conical flask. Clean the pipette by sucking up the analyte solution and wetting the surface inside by tilting and rotating it. The 'rinse' solution drains through the pipette tip and is disposed of.

Fill the pipette with the analyte solution to above the graduation (scratch) mark. Allow the solution to drain slowly from the vertical pipette until the bottom of the meniscus is level with the graduation mark.

Place the pipette tip in the neck of the conical flask and allow the analyte solution to drain. After it stops flowing, touch the tip against the inside wall of the flask for 30 seconds. Then add a few drops of an indicator to the analyte solution in the conical flask.

Place the conical flask containing the analyte solution and indicator under the burette, making sure that the tip of the burette is inside the neck of the conical flask. A white ceramic tile underneath the flask will allow you to see the colour change at the end-point more clearly (Figure 2.4).

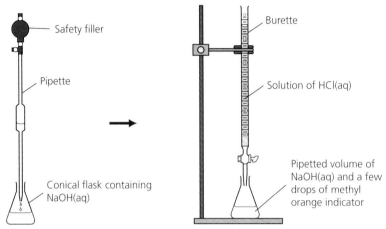

Figure 2.4 Apparatus for a titration

Examiner guidance

The first titration is a **trial run** to see the colour change of the indicator and to find an approximate value of the titre volume. Quickly add small volumes of the standard solution from the burette. You must swirl the conical flask to help the mixing process and give the reacting species in solution time to react. Continue doing this until the end-point is reached and then record the final burette reading.

Complete the titration by adding the standard solution very slowly, drop-wise, while swirling the conical flask. When the indicator just changes colour, you have reached the end-point of the titration. Record the final burette reading to 2 decimal places: the last number will be a 0 or 5.

The titre results of one person performing the same titration (or another experiment) many times or of many students repeating the same titration should fall into a normal or Gaussian distribution curve (Figure 2.5).

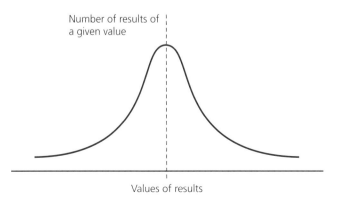

Figure 2.5 Normal or Gaussian distribution curve

Expert tip

If any of the solution splashes onto the walls of the flask then wash it into the mixture with distilled water. If you are close to the end-point and there is a drop of the solution stuck to the tip of the burette, remove it by touching the tip to the wall of the flask and washing it into the solution.

RESOURCES

The Royal Society of Chemistry provides additional information on how to set up experiments:

http://www.rsc.org/Education/ Teachers/Resources/practical/ index3.htm

It is difficult to determine accurately the volume of solution in a burette if the meniscus lies between two graduation marks.

The chemical reagent used to prepare a standard solution may not be 100 % pure.

A 250 cm³ volumetric flask may actually contain 250.3 cm³ when filled to the calibration mark due to permitted variation in the manufacture of the flask.

A burette is calibrated by the manufacturer for use at 20 °C. When it is used in the laboratory the temperature may be 23 °C. This difference in temperature will cause a small difference in the actual volume of solution in the burette when it is filled to a calibration mark.

It is difficult to make an exact judgement of the end-point of a titration (the exact point at which the colour of the indicator changes).

The display on a laboratory balance will only show the mass to a certain number of decimal places.

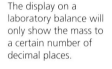

Expert tip

To minimize experimental errors (Figure 2.6) or increase reliability during titrations:

- it is important to swirl the mixture during titration as this ensures that the reactants are well mixed
- repeat your titration until you obtain at least two consistent results (two titre volumes that are within ±0.10 cm³ of each other); replication involves repeating the experiment a number of times
- always repeat the experiment to check your results are reliable; calculate an average value to minimize random errors
- treat burettes with a cleaning mixture to remove grease, so they drain regularly
- prepare your sodium hydroxide solution freshly, since it absorbs carbon dioxide from the air which reduces its concentration.

Figure 2.6 Sources of uncertainty in volumetric analysis

Indicators

Choosing an indicator (Chapter 1) for a titration depends on the type of acid–base reaction taking place (Table 2.2).

Acid–base reaction type	pH at equivalence point	Suitable indicator
Strong acid/strong base	7	Any
Weak acid/strong base	>7	Phenolphthalein
Strong acid/weak base	<7	Methyl orange
Weak acid/weak base	~7	None

Table 2.2 Types of acid–base titration and choice of indicator

Key definition

Acid–base indicator – an indicator that changes colour on going from acidic to basic solutions.

Expert tip

It is important to use only a few drops (2 or 3) of indicator. This is because indicators are weak acids and so are neutralized by alkalis.

ACTIVITIES

3 Find out about the double indicator method.

4 Distinguish between the terms **end-point** and **equivalence point**.

Types of titration

There are three main types of titration:

- **Acid–base titrations**, in which protons are transferred from the acid to the base.

- **Redox titrations**, in which an oxidizing agent is titrated against a reducing agent (or *vice versa*) usually in aqueous acidic solution. A redox titration may use a redox indicator that has one colour in its reduced state and a different colour in its oxidized state.

- **Complexometric titrations**, which are based on complex formation involving a Lewis acid–base reaction between metal ions and ligands, in which the ligands use their lone pairs of electrons to form coordinate bonds with metal ions.

The most common ligand (complexing agent) used is ethylenediaminetetraacetic acid (EDTA). It is a hexadentate ligand that forms very stable 1:1 complexes with metal ions (regardless of charge):

$$M^{n+} + EDTA^{4-} \rightarrow MEDTA^{n-4}$$

Most titrations are direct. 'Direct' means one reagent is added directly to the other until the end-point is reached. If a direct titration is not possible, a back titration may be used.

Back titrations

A back titration involves adding a known but excess amount of one standard reagent to a known mass of the substance being determined (the analyte). After the reaction between the two is complete, the excess amount of the standard reagent is determined by titration against a second solution of a primary standard.

Back titrations are used when:

- no suitable indicator is available for a direct titration

- the end-point of the back titration is clearer than that of the direct titration

- the reaction between the standard reagent and analyte is slow

- the analyte is insoluble in water.

Units of concentration

The SI unit for concentration is moles per cubic decimetre ($mol\,dm^{-3}$) but there are a number of alternatives for expressing the relative amounts of solute and solvent that you may encounter during practical work or in the chemical literature.

Molality

This is used to express the concentration of solute relative to the mass of the solvent. It has SI units of $mol\,kg^{-1}$. It is a temperature-independent measure of concentration and is used when the osmotic properties of the solution are relevant.

Percent composition (% w/w)

This is the solute mass (in g) per 100 g of solution. The advantage is that a solution can be prepared easily by pre-weighing the solvent and solute and then mixing.

> **Expert tip**
>
> A complexometric back titration is needed if:
>
> - the metal ion precipitates in the absence of EDTA
> - the metal ion reacts very slowly
> - the metal ion forms an inert complex
> - no suitable indicator is available.
>
> In a complexometric back titration, a known excess of EDTA is added to the metal ion (buffered to an appropriate pH). Then the excess EDTA is titrated with a standard solution of a different metal ion.

> **Worked example**
>
> Calculate the concentration in % (w/w) for a 1.00 M solution of sodium chloride, NaCl(aq).
>
> $$\text{Concentration} = \frac{1\,mol}{1\,dm^3} = \frac{58.44\,g}{1\,dm^3} = \frac{5.844\,g}{100\,cm^3} = 5.8\,\%\ (w/w).$$

Percent concentration (% w/v and % v/v)

This is the solute mass (in g) per 100 cm^3 of solution. This is more commonly used than percent composition, since solutions can be accurately prepared by weighing out the required mass of solute and then making this up to a known volume using a volumetric flask. The equivalent expression for liquid solutes is % v/v.

Parts per million (ppm)

This is a non-SI weight per volume (w/v) concentration term commonly used to describe very low concentrations of chemicals. The term ppm is equivalent to the expression $\mu g\,cm^{-3}$ ($10^{-6}\,g\,cm^{-3}$) and a 1.0 ppm solution of a substance will have a concentration of $1.0\,\mu g\,cm^{-3}$.

■ ACTIVITIES

5 250 cm^3 of water contains 2.2 mg of dissolved oxygen. Determine the concentration in ppm.

6 The molarity of white vinegar ($CH_3COOH(aq)$) is 0.8393 mol dm^{-3}. The density of white vinegar is 1.006 g cm^{-3}. The molar mass of ethanoic acid is 60.05 g mol^{-1}. Calculate the mass percent composition (w/w) of white vinegar.

7 Find out about the use of 'volume strength' as a measure of the concentration of hydrogen peroxide.

Redox titrations with starch and iodine

The starch solution is added to the reagent in the flask at the start of the titration and the end-point is from colourless to blue. If the iodine solution is being titrated, the starch must be added later because the concentration of iodine would be so high that some of the molecules would interact permanently with the starch and not be free to react with the titrant.

The starch is therefore added once most of the iodine molecules have been reduced; that is, when the initial brown colour of the solution has faded to a very pale yellow colour. On adding the starch, the solution turns blue. The titration is complete when the blue colour disappears.

■ ACTIVITIES

Some iron tablets, containing iron(II) sulfate, were left open on the shelf so that some of the iron(II) sulfate was oxidized into iron(III) sulfate.

You are to analyze by titration the percentage by mass of iron(II) ions present. The partially oxidized iron tablets are dissolved in water to release the iron(II) and iron(III) ions. A suitable volume of dilute sulfuric acid is then added to the solution to prevent oxidation of iron(II) to iron(III) ions.

An unoxidized sample of an iron tablet is dissolved and made up to 250.00 cm^3. 25.00 cm^3 of the iron(II) ions in solution required 14.00 cm^3 of potassium dichromate(VI) to reach an end-point in a titration.

8 Calculate the mass of iron(II) sulfate in the unoxidized sample of iron tablet that was used.

Ionic reaction:

$6Fe^{2+} + Cr_2O_7^{2-} + 14H^+ \rightarrow 6Fe^{3+} + 2Cr^{3+} + 7H_2O$

9 Outline how the iron(III) ions present in the solution formed from a dissolved iron tablet can be converted to iron(II) ions and then analyzed by a redox titration. Zinc powder is available in the laboratory.

10 Outline how that procedure could be modified, using excess powdered pure zinc, to determine the amount of iron(III) ions in an oxidized iron tablet sample.

Ideas for investigations involving titration

- Analyze wines for ethanol concentration (dichromate(vi)) and iodine and starch.
- Investigate nitrification in filtered soil solution via back titration.
- Analyze the vitamin C content of fruit juices (using DCPIP, iodine or iodate(v) ions).
- Investigate the titration of decarbonated fizzy drinks using sodium hydroxide and phenolphthalein.
- Determine the chloride ions in cheese using the Volhard method.
- Investigate calcium and magnesium ion concentration (using EDTA in the presence of eriochrome black T indicator).
- Investigate the copper and zinc content of brass using redox titrations.
- Investigate purity and composition of antacids by back titration.
- Determine the concentration of bleach (sodium chlorate(i) solution) by redox titration.
- Investigate the reaction between bromine and thiosulfate ions via a redox titration.

Gravimetric analysis

Gravimetric analysis involves the accurate measurement of the mass of a product from an accurately measured mass of a reactant. There are two types: volatilization and precipitation.

In a volatilization method, you weigh out a sample of the analyte and heat it to constant mass using a crucible (Figure 2.7). You can then collect the product and weigh it, or determine its mass indirectly from its loss in mass. You can use a desiccator (page 19) to store the sample while it is cooling down.

In a precipitation method, you dissolve the analyte in water and convert it into an insoluble product by reacting it with another chemical (reagent). Then filter, wash, dry and weigh the precipitate.

In precipitation methods, the particle size of the solids should be large because large particles are easier to filter. Small particles could block the filter paper or pass through it. A solid that consists of large particles has a smaller total surface area and so can be washed free of impurities more easily.

Stir rapidly and heat the reaction mixture on a steam bath.

Key definition

Precipitation reaction – a reaction that involves the formation of an insoluble salt when two solutions containing soluble salts are combined. The insoluble salt formed is known as the precipitate.

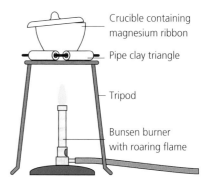

Figure 2.7 Gravimetric analysis using a crucible to heat magnesium to constant mass

Crucible containing magnesium ribbon
Pipe clay triangle
Tripod
Bunsen burner with roaring flame

Expert tip

To minimize experimental errors or increase reliability for gravimetric analysis:

- heat the solid evenly to make sure all of the solid has undergone decomposition
- cool the heated sample in a desiccator to prevent the absorption of water vapour from the air
- repeat the heating–cooling–weighing cycle until constant mass occurs
- ensure all the ions are precipitated out during a precipitation gravimetric analysis by using excess precipitating reagent
- wash the filtered residue with a suitable solvent that will remove impurities, but does not dissolve the residue
- plan to have sufficient precipitate so the effect of random uncertainty in the measurements is relatively low.

■ ACTIVITY

11 Magnesium sulfate does not decompose when heated strongly with a
   Bunsen burner. However, magnesium carbonate decomposes on heating to
   form magnesium oxide and carbon dioxide.

   You are to design an experiment to determine the percentage by mass of
   magnesium carbonate in a sample of magnesium carbonate contaminated
   by magnesium sulfate.

   Outline the practical sequence for the method to record appropriate masses,
   decompose magnesium carbonate in the sample by heating and ensure that
   thermal decomposition is complete. Show how you would tabulate your results.

Ideas for investigations involving gravimetric analysis

- Determine chloride ion concentration in sea water or foods (using silver
  nitrate solution).
- Analyze water of crystallization in hydrated salts.
- Analyze mixtures of two salts; analyze impure sodium thiosulfate.
- Investigate the reaction between copper(ɪɪ) sulfate and excess sodium
  carbonate solution.

Colorimetric and spectrophotometric analysis

Colorimetric analysis is used to determine the concentration of analytes that
are coloured or can be converted quantitatively by a chemical reaction into a
coloured species.

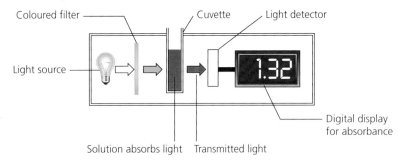

Figure 2.8 The pathway of visible light in a colorimeter (a blue or green
filter is used with a red solution)

The colour of a solution depends on the frequencies/wavelengths of light it absorbs
(and this in turn determines the wavelengths of light that are transmitted). The
intensity of its colour depends on the concentration of the solution: the more
concentrated the solution, the darker its colour, as it absorbs more light.

Examiner guidance

Generally speaking, there is a linear relationship between the concentration
of a solution and the absorbance (known as the **Beer–Lambert law**.
However, high absorbance values above 2 will generally not obey this linear
relationship. If readings of over 2 are obtained, dilute the sample by a known
amount and re-measure the absorbance.

A standard calibration line often has to be obtained; this involves preparing a
series of solutions of known concentration by the accurate dilution of a standard
solution. The absorbance values of these standard solutions are measured using the
instrument (with the correct filter or correct wavelength, for maximum absorption).

Key definition

Beer–Lambert law – the linear
relationship between absorbance
and concentration of an absorbing
species.

■ **ACTIVITY**

12 Plot a calibration line for the following data for aqueous solutions of chromium(III) nitrate and use it (via **interpolation**) to calculate the concentration of a solution with an absorbance of 0.256.

Solution number	Concentration/mol dm⁻³	Absorbance
1	0.020	0.333
2	0.010	0.163
3	0.005	0.084
4	0.0025	0.041
5	0.0013	0.019
6	0.00065	0.011

Ideas for investigations involving colorimetric or spectrophotometric analysis
• Investigate the movement of coloured transition metal ions through a semi-permeable membrane.
• Determine the concentration of lactose in milk via colorimetry using bicinchoninic acid.
• Determine the formula of a hydrated transition metal ion complex.
• Investigate the extraction of iron or manganese from tea by oxidation of cations followed by colorimetric analysis.
• Investigate the reaction between copper(II) ions and aspirin via colorimetry in the presence of potassium hydrogencarbonate.

Many laboratory procedures, including titrations and spectrophotometry, may involve dilutions.

Dilutions

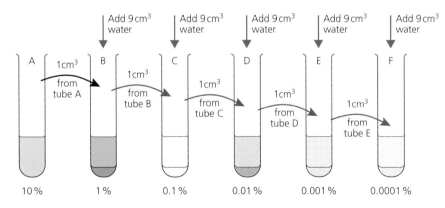

Figure 2.9 Making a serial dilution: dilution factor of 10 for each transfer

> **Key definition**
>
> **Serial dilution** (Figure 2.9) – any dilution where the concentration decreases by the same factor in each successive step.

For example, a $\frac{1}{10}$ dilution of a solution represents 1 part of the chemical solution added to 9 parts of diluent (usually water). Serial dilutions are multiplicative. For example, if a $\frac{1}{8}$ dilution of the chemical stock solution is made followed by a $\frac{1}{6}$ dilution with water, then the final dilution is $\frac{1}{8} \times \frac{1}{6} = \frac{1}{48}$.

Doubling dilutions are a series of ½ dilutions: each successive solution will contain half the amount of the original concentrated (stock) solution. Table 2.3 shows the effect of doubling dilutions performed six times. This results in a series of dilutions, each a doubling dilution of the previous one.

Dilution	Concentration as a fraction of the original
First dilution	$\dfrac{1}{2}$
Second dilution	$\dfrac{1}{2} \times \dfrac{1}{2} = \dfrac{1}{4}$
Third dilution	$\dfrac{1}{2} \times \dfrac{1}{2} \times \dfrac{1}{2} = \dfrac{1}{8}$
Fourth dilution	$\dfrac{1}{2} \times \dfrac{1}{2} \times \dfrac{1}{2} \times \dfrac{1}{2} = \dfrac{1}{16}$
Fifth dilution	$\dfrac{1}{2} \times \dfrac{1}{2} \times \dfrac{1}{2} \times \dfrac{1}{2} \times \dfrac{1}{2} = \dfrac{1}{32}$
Sixth dilution	$\dfrac{1}{2} \times \dfrac{1}{2} \times \dfrac{1}{2} \times \dfrac{1}{2} \times \dfrac{1}{2} \times \dfrac{1}{2} = \dfrac{1}{64}$

Table 2.3 Series of dilutions

The dilution factor is calculated from the formula: $\dfrac{\text{final volume}}{\text{aliquot volume}}$.

Worked example

Deduce the dilution factor if you add a $0.1\,cm^3$ aliquot of a specimen to $9.9\,cm^3$ of diluent.

The final volume is equal to the aliquot volume plus the diluent volume:

$0.1\,cm^3 + 9.9\,cm^3 = 10\,cm^3$

The dilution factor is equal to the final volume divided by the aliquot volume:

$\dfrac{10\,cm^3}{0.1\,cm^3} = 100 : 1$ dilution

Chromatography

All types of chromatography involve a stationary phase (usually silica, SiO_2, or cellulose) and a mobile phase (usually a solvent system). Two compounds usually have different partitioning characteristics between the stationary and mobile phases. Since the mobile phase is moving, then the longer the time a compound spends in that phase, the further it will travel.

Analytical techniques may be used to follow the course of reactions and determine the purity of products. These methods include GC, HPLC and TLC. Sample sizes are usually from microgram to milligram quantities.

Preparative methods are used to purify and isolate organic compounds for characterization or further use. The common techniques are preparative HPLC, preparative TLC and column chromatography.

Ideas for investigations

Ideas for possible investigations involving paper chromatography include:

- investigating the presence of different chloroplast pigments
- studying pigments in red and green peppers
- studying hydrolyzed proteins (using locating agents)
- studying inorganic ions and inorganic precipitates.

Possible investigations involving TLC include:

- monitoring the esterification of benzene carboxylic acid
- separating phenols
- investigating the products of the nitration of phenol
- isolating organochlorine pesticide residues.

Expert tip

Sometimes a single solvent is used in preparative chromatography, but usually it is a binary mixture of solvents with different polarities. The advantage is that the bulk polarity can be controlled by varying the ratio of the two solvents.

Practical organic chemistry

Practical organic chemistry often involves extracting a natural product, or synthesizing a pure sample of a specific organic compound that involves the following steps:

- **preparation**: carry out the reaction or series of reactions and prepare a crude (impure) sample of the organic product

- **isolation**: separate the crude product from the reaction mixture

- **purification**: remove impurities (work-up) from the crude product

- **identification**: confirm the identity of the pure product

- **calculation**: calculate the percentage yield (and atom economy).

> **Examiner guidance**
>
> You must carry out a risk assessment before undertaking any practical work with organic chemicals. You must show you are aware of all risk and hazards in your planned practical work.

▧ Preparation

Organic preparations are often carried out using Quickfit® glassware with ground glass joints that allow the individual pieces to fit together tightly, so no corks or rubber stoppers are needed.

Figure 3.1 shows a round-bottomed flask and a condenser used for heating liquids under reflux. An electric heating mantle is used to heat the reactants. It contains a cavity, shaped to hold the reaction flask (about half full) so its heating can be controlled.

> **Examiner guidance**
>
> Organic compounds are generally flammable so, to prevent fires, there must be no naked flames (Bunsen burners) around.

> **Expert tip**
>
> Heating may also involve using a hot-plate with a magnetic stirrer. Remember that a hot surface can be a source of ignition. A hot plate can retain its heat and it is easy to burn your hand/skin on it. You must **not** add solvent to a hot vessel.

You need to know the mass of the limiting organic reactant so you can calculate the theoretical yield and percentage yield of the organic product. The other reactants can be added with anti-bumping granules to stop chemicals 'bumping' and leaving the flask. If any of the reactants are solids or immiscible liquids, you might need a solvent to produce a homogeneous mixture.

On heating, the more volatile (lower boiling point) chemical will boil and its gas will enter the condenser. It will be cooled by the cold water running through the outer tube of the condenser, so it will condense and return to the reaction flask. The condenser prevents the escape of any volatile organic chemicals. The process of boiling a reaction mixture and condensing the hot gases back into the reaction flask is called **heating under reflux**. A drying tube may be placed on top of the condenser.

> **Expert tip**
>
> The clamps should support apparatus from the bottom upwards and not too many clamps should be used: they can introduce stresses. Dirty joints can stick and so you should avoid getting chemicals on the joints. You should also clean the apparatus immediately after use. You need to check that the joints are secure without being stuck.

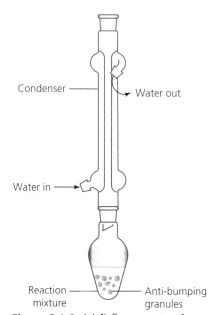

Condenser — Water out

Water in —

Reaction mixture — Anti-bumping granules

Figure 3.1 Quickfit® apparatus for heating under reflux

> **Expert tip**
>
> You must always add anti-bumping granules **before** heating begins, because adding them to a hot mixture is likely to cause it to bubble over.

Isolation

After the preparation stage of a synthesis experiment, there will be a mixture of organic and inorganic substances (usually ionic) in the reaction flask. In addition to the organic product, the mixture is likely to contain:

- reactants that were used in excess

- other products of the reaction

- compounds that are produced as a result of side reactions

- the limiting reactant if the reaction was reversible.

Filtration

The next step is to separate the organic compound from the other chemicals in the mixture. If the product is a solid, then it can be removed by filtration under reduced pressure or vacuum filtration. This type of fast filtration is carried out using a Büchner funnel and Büchner flask (Figure 3.2).

The Büchner funnel has a plate with a number of small holes. The Büchner flask has thick walls (to withstand changes in air pressure) with a side arm. The funnel is fitted into the neck of the flask by a rubber stopper and the flask is attached to a pump via its side arm. The Büchner flask and funnel will also need a rubber ring at the neck of the flask (to ensure an excellent vacuum tight seal).

Place a filter paper flat on the perforated plate in the funnel and wet it. The suction from the pump will make the filter paper stick to the perforated plate; this will stop any solid from passing round and under the edge of the filter paper into the flask. The Büchner flask should be clamped since it is top-heavy.

The product is then washed with a liquid, often ice-cold water, to remove any impurities sticking to its surface. The crude organic product is partially dried by drawing air through it.

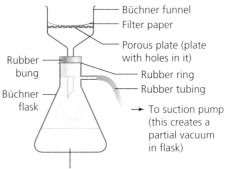

Filtrate (liquid that passes through filter paper) collects here

Figure 3.2 Suction filtration

Simple distillation

If the organic product is a liquid in the reaction mixture and it is more volatile than the other substances in the mixture, then you can remove it by simple distillation (Figure 3.3). When a liquid boils, the boiling point is constant and the temperature remains constant. You can add boiling chips to promote smooth boiling.

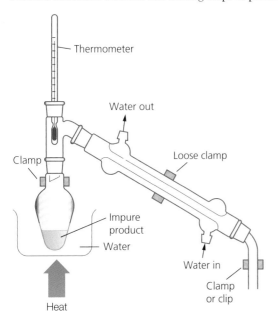

Figure 3.3 The Quickfit® glassware for simple distillation

■ ACTIVITIES

1 Explain why the thermometer is placed at the entrance where the vapour exits into the Liebig condenser.

2 Explain why the cooling water is run into the Liebig condenser at the bottom rather than at the top.

■ Fractional distillation

Fractional distillation is used if there are more than two liquids, or if the two boiling points are close. The liquids with the higher boiling points will condense at the bottom of the fractionating column (which is filled with glass beads for the vapour to condense on) and fall back into the flask again. The liquids that have lower boiling points will condense higher up the column. At each point of the fractionating column, the temperature corresponds to the boiling point of a particular liquid, which means that many simple distillations take place at different points of the column.

▨ Purification

■ Recrystallization

Recrystallization is used if the organic product is a crystalline solid. You dissolve the organic compound and impurities in a small volume of a solvent at a higher temperature. The term 'recrystallization' is used because it involves dissolving a solid that had originally crystallized from a reaction mixture and causing it to crystallize again from solution.

With a higher solubility at the higher temperature, more of the organic compound can be dissolved in a small volume of solvent. This means that when the solution is slowly cooled down, the organic compound crystallizes out, leaving the impurities behind in the solvent (Figure 3.4).

Expert tip

An important and common method for concentrating substances is rotary evaporation. The solvent is removed under reduced pressure while the flask containing the substance is rotated in a water bath. The temperature of the water bath is controlled and the vaporized solvent is collected in a separate flask after condensing.

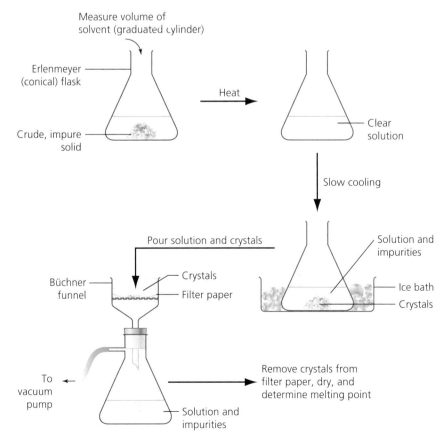

Figure 3.4 The principle of recrystallization

The hot mixture is then filtered to keep the residue of the organic compound. During recrystallization you should use a fluted filter paper (to maximize surface area) to filter the hot solution, in order to remove insoluble impurities.

However, if the impurities are insoluble in the solvent, you can dissolve the mixture using the hot solvent so it passes through the filter paper. The dissolved compound is collected, leaving the impurities in the filter funnel. Cool the filtrate down slowly to favour the formation of pure crystals. Rapid cooling may trap impurities inside the crystals.

Expert tip

You can induce crystallization by using an ice bath, adding dry ice or scratching the inside of the conical flask to provide tiny glass particles as nucleation sites.

Examiner guidance

Various forms of chromatography (see Chapter 2) are also used in organic chemistry to separate and identify organic compounds. If a substance refuses to crystallize, it may be very impure so you should use chromatography to investigate its composition.

■ ACTIVITY

3   Find out about **oiling out** and how to deal with it.

■ Solvent extraction

It may not be practical to separate the organic product directly from the mixture by filtration or simple distillation, so solvent extraction is used (Figure 3.5).

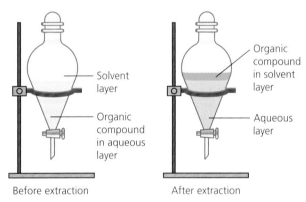

Before extraction — Solvent layer / Organic compound in aqueous layer

After extraction — Organic compound in solvent layer / Aqueous layer

Figure 3.5 Solvent extraction

If the organic product is present in an aqueous mixture, it can be removed by adding a second solvent (for example, dichloromethane, CCl_2H_2). This second solvent must be immiscible with water, so on mixing they form separate layers. The product must not react with the solvent and it must be more soluble in it than water ('like dissolves like' principle). When the solvent is added to the aqueous mixture the molecules of the organic product will move into the organic solvent layer, and so can be separated.

The aqueous layer is transferred to a separating funnel and a sample of solvent (less than half of the volume of the aqueous mixture) is then added. The stoppered funnel is inverted and the tap opened to release any pressure from the solvent vaporizing. The tap is closed and the mixture is shaken. The funnel should be inverted several times. This increases the surface area of contact between the two liquids, which increases the rate of movement of the organic product into the solvent layer. The organic extract can be dried over anhydrous sodium sulfate before evaporating the solvent to recover the compound.

Expert tip

When you are carrying out a separation you should note which phase is the organic solvent and which is the water. For example, dichloromethane will be the lower while ethoxyethane/ethyl ethanoate will be the upper. If you have carried out a reaction in ethanol, this will make an organic solvent miscible with water.

Expert tip

Multiple extractions with small amounts of solvent are more efficient than one single extraction with a large volume. This is a direct result of the distribution coefficient.

Examiner guidance

Organic acids (sulfonic acids and carboxylic acids) are easily converted into their soluble sodium salts by reaction with sodium hydrogencarbonate. Phenols are weaker acids and require sodium hydroxide. Bases, such as amines, are converted into soluble salts by reaction with dilute hydrochloric acid.

■ ACTIVITIES

In an experiment to prepare cyclohexene, C_6H_{10}, concentrated sulfuric acid was added drop by drop to 6.818 cm³ of cyclohexanol ($C_6H_{11}OH$, density 811 kg m⁻³), refluxed and then distilled.

As the reaction took place, cyclohexene distilled into the collection flask (receiver) and a black deposit of carbon (soot) formed in the reaction flask. This is known as 'charring'.

The equation for the main dehydration reaction is:

$$C_6H_{11}OH(l) \rightarrow C_6H_{10}(l) + H_2O(l)$$

4   After purification, 1.85 g of cyclohexene was collected. Calculate the amount (number of moles) of cyclohexanol used in the experiment.

5   Calculate the mass of cyclohexene that should be formed (theoretical yield) if all of the cyclohexanol is converted into cyclohexene.

6   Calculate the percentage yield of cyclohexene.

7   Explain the importance of the different boiling points of cyclohexanol (161 °C) and cyclohexene (83 °C) to the success of the synthesis.

8   Explain why the formation of carbon (via a charring process) reduces the yield of cyclohexene.

9   With reference to the use of concentrated sulfuric acid in the synthesis, suggest what safety precautions should be taken when clearing up the apparatus after the preparation.

10 Identify a chemical reagent that could be used to test for the presence of the alkene double bond (unsaturation) in cyclohexene. Describe the expected result of the test.

▨ Identification

If the product is a solid then you can determine its melting point and compare this with the **literature value**. If the two values are similar then the compound is present and relatively pure.

The melting point of a solid is the temperature at which it changes into a liquid (at constant temperature). The melting point can be determined using the apparatus shown in Figure 3.6. Both sets of apparatus measure a melting range. A fine dried powdered sample of the purified organic compound is needed.

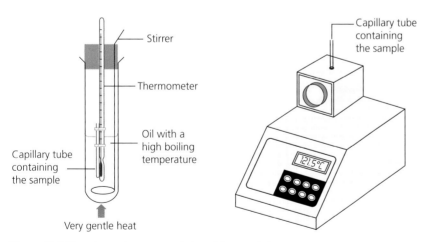

Figure 3.6 Simple and commercial melting point apparatus

You can use a variety of spectroscopic techniques, such as IR spectroscopy, UV-vis spectroscopy and nuclear magnetic resonance (NMR), to establish or confirm the structure of the organic product.

The presence of impurities in an organic substance **lowers** its melting point and **widens** its melting point range. The greater the amount of impurity present, the greater the lowering of the melting point. This is because the impurity molecules disrupt the regular packing in the lattice of the compound.

■ ACTIVITIES

11 N-phenylethanamide has a literature melting point of 114.3 °C. A sample you have synthesized has a melting range of 107 to 110 °C. Deduce three possible conclusions from this result.

12 Find out about the method of mixed melting points.

13 Find out how **Brady's reagent** is used to identify carbonyl compounds.

If the organic product is a liquid then you can identify it by its boiling point via simple distillation. During distillation the temperature that remains constant is the boiling point of the organic liquid. However, boiling points vary with pressure, so liquids are often converted to solid derivatives because melting points are not affected by pressure.

Organic reports

For an organic synthesis report, the main reaction and its mechanism should be shown using the arrow notation (see Chapter 10). In your report you should describe side reactions and how they are minimized, then give a description of the purification, explaining how the main organic product is separated from the side products, unreacted reagents, catalysts and solvents.

Data tables should be included for the reactants (reagents) and products, as shown in Table 3.1. State the boiling point and density for liquids, but only the melting point for solids. List only reagents, not solvents or catalysts.

Compound	Molar mass /g mol^{-1}	Mass used/g	Amount used/mol	Melting point/°C	Boiling point/°C	Density

Table 3.1 Outline data table for organic synthesis

The results of the experiment should include: the experimental yield (as a mass, amount and percentage), physical state, appearance, melting or boiling point, results of any chemical or physical tests and the results of any chromatographic or spectroscopic analysis.

You should discuss whether the organic compound was obtained reasonably pure and in good yield, with a summary of the evidence that supports that conclusion. The discussion should include an error analysis giving reasons for low yield and possible contaminants. You should also include a summary of evidence supporting the identity of the product.

Ideas for organic investigation

- Synthesize pharmaceuticals (for example, aspirin and paracetamol (acetaminophen)).
- Synthesize dyes (for example, indigo).
- Synthesize photochromic organic compounds.
- Synthesize insect pheromones.
- Synthesize esters.
- Synthesize substituted aromatic compounds.
- Isolate and study caffeine from tea.
- Isolate and study eugenol from cloves.
- Isolate and study lycopene from tomatoes.
- Isolate and study limonene from citrus fruits.
- Isolate and study theobromine from cocoa.
- Investigate reaction mechanisms (for example, hydrolysis of halogenoalkanes).
- Investigate novel reactions, such as the Cannizzaro or Diels–Alder reaction.

Molecular modelling kits (mandatory practical)

There are a number of commercially available molecule building kits. These are often colour coded: H (white; 1 hole), O (red; 2 holes), C (black; 4 holes). Grey links represent bonds: short links represent single bonds and long links represent double or triple bonds.

Worked example

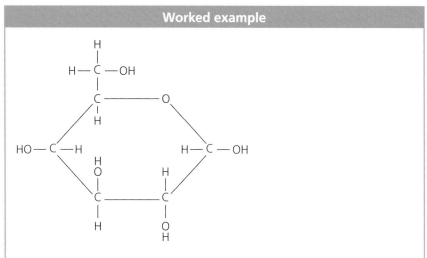

Figure 3.7 The molecular structure of glucose

Assemble the model by making a chain of five carbon atoms. Add an oxygen atom to the chain of carbons, and then form a ring (which is puckered). Build $-CH_2OH$ and add this on to the carbon ring, then form four hydroxyl groups ($-OH$). Add the hydrogen atoms to the remaining holes.

4 Mathematical and measurement skills

Scientific notation and significant figures

Scientific notation

Very large and very small numbers are often written in **scientific notation**: $N \times 10^n$, where N is a number between 1 and 10 and n is the exponent (power) to which 10 is raised.

ACTIVITIES

1 Write the following numbers in scientific notation:

   1002   54   6 926 300 000   −393   0.003 61   −0.0038

2 Write the following numbers in ordinary notation:

   1.93×10^3 3.052×10^1 -4.29×10^2 6.261×10^6 9.513×10^{-8}

Concept of significant figures

The number of **significant figures** in a number indicates the error in a measurement (Figure 4.1).

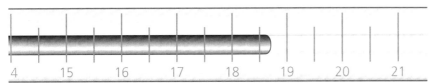

Figure 4.1 A magnified thermometer scale showing a temperature of 18.7 °C: the last digit is uncertain

The following rules should be applied to establish the number of significant figures:

- Zeros between digits are significant. For example, 2006 has four significant figures.

- Zeros to the left of the first non-zero digit are not significant (even when there is a decimal point in the number). For example, 0.005 has one significant figure.

- When a number with a decimal point ends in zeros to the right of the decimal point these zeros are significant. For example 2.0050 has five significant figures.

- When a number with no decimal point ends in several zeros, these zeros may or may not be significant. The number of significant figures should then be stated. For example, 30 000 (to 3 significant figures) means that the number has been measured to the nearest 100.

Examiner guidance

Place-holder zeros can be removed by converting numbers to scientific notation. For example, 2000 may have one significant figure, but 2.00×10^3 has 3 significant figures.

When significant figures are used as an implicit way of indicating uncertainty, the last digit is considered uncertain. For example, a result reported as 1.23 implies a minimum uncertainty of ±0.01 and a range of 1.22 to 1.24.

■ ACTIVITY

3   State and explain the number of significant figures in the following
measurements:

14.44   9000   3000.0   1.046   0.26 6.02×10^{23}

■ Rounding off significant figures

A digit of 5 or more rounds up; a digit smaller than 5 rounds down. When rounding,
look only at the one figure beyond the number of figures to which you are rounding,
so to round to three significant figures, only look at the fourth figure.

> ### Worked example
>
> The number 350.99 is 351.0 rounded to 4 significant figures; 351 to 3 significant
> figures; 350 to 2 significant figures and 400 to 1 significant figure.

■ ACTIVITY

4   Report the following numbers to three significant figures:

654.389   65.4389   654 389   56.7688   0.035 422 10

■ Significant figures in simple calculations

A calculated result from measurements must be reported with the correct number
of significant figures. This depends not only on the number of significant figures
in the individual measurements, but also on the type of mathematical operation
used to obtain the result.

When multiplying and dividing measurements, the result must have the same
number of significant figures as the measurement with the fewest significant
figures.

> **Common mistake**
>
> Do not simply copy down the
> final answer from the display of
> a calculator. This often has far
> more significant figures than the
> measurements justify and you will
> lose marks for this under the Analysis
> and Communication Criteria.

> ### Worked example
>
> Calculate the amount of solute in $0.025\,30\,dm^3$ of an aqueous solution with
> concentration $1.20\,mol\,dm^{-3}$.
>
> Amount of solute = concentration × volume
>
> $$= 1.20\,mol\,dm^{-3} \times 0.025\,30\,dm^3$$
>
> $$= 0.030\,36\,mol \text{ (calculator answer)}$$
>
> The concentration measurement has the lower number of significant figures, in
> this case three, and so the result must be quoted to 3 significant figures (sf). The
> calculator answer is therefore rounded: amount of solute = 0.0304 mol.

When adding and subtracting measurements, the result must be reported with the
same number of decimal places as the measurement with the lowest number of
decimal places.

> ### Worked example
>
> A solution was prepared by dissolving 2.55 g of the solid in 26.2 g of water.
> Calculate the total mass of the solution.
>
> Total mass = 28.75 g. The measurement with the lower number of decimal
> places is the 26.2 g of water and so the result must be quoted to 1 decimal
> place: total mass = 28.8 g.

■ ACTIVITIES

Solve the following calculations and give the answers with the correct number of significant figures:

5 6.201 + 7.4 + 0.68 + 12.0

6 1.6 + 1.62 + 1200

7 8.264 – 7.8

8 10.4168 – 6.0

9 1.31 × 2.3

10 5.7621 × 6.201

11 20.2 ÷ 7.41

12 40.002 ÷ 13.000 005

For multiple calculations, determine the number of significant digits to keep in the same order as the operations: first logarithms and exponents, then multiplication and division, and finally addition and subtraction. When parentheses are used, do the operations inside the parentheses first. To avoid rounding errors, keep extra digits until the final step.

■ Mathematical and physical constants

When using a mathematical formula which contains constants such as π or e, the value contains as many significant figures as you enter in the formula. When using a mathematical formula which contains integers, the integers are assumed to have infinite significance, and do not limit the number of significant figures in the result.

■ Introducing rounding errors in multi-step calculations

Some calculations involve multiple steps, such as calculating enthalpy changes. Care must be taken to avoid rounding errors, as shown in the calculations below.

When 6.074 g of a metal carbonate is reacted with 50.00 cm^3 of 2.00 mol dm^{-3} hydrochloric acid (present in excess), a temperature rise of 5.4 °C is obtained. The specific heat capacity of the solution is 4.18 J g^{-1} °C^{-1}. (The density of the solution is assumed to be 1.00 g cm^{-3}.)

Heat produced = 50.00 g × 4.18 J g^{-1} °C^{-1} × 5.4 °C = 1128.6 J = 1.1286 kJ (by calculator)

Since the least precise measurement (the temperature rise) only has 2 sf, the energy produced should have 2 sf. Therefore the heat released should be quoted as 1.1 kJ. If this figure is used to calculate the enthalpy change per mole, however, a 'rounding error' will be produced. The 1.1286 kJ value must be used.

If the metal carbonate has a molar mass of 84.32 g mol^{-1}, the enthalpy change per mole of the metal carbonate can be calculated from the value above.

Using the calculator value of 1.1286 kJ for the heat produced:
    enthalpy per mole = 15.667 361 29 kJ mol^{-1} = 16 kJ mol^{-1} (2 sf)

Using the rounded value of 1.1 kJ for the energy produced:
    enthalpy per mole = 15.270 332 64 kJ mol^{-1} = 15 kJ mol^{-1} (2 sf)

Hence, rounding to 2 sf too early produces a 'rounding error'.

■ Logarithms and significant figures

The number of significant figures in the mantissa of log x is the same as the number of significant figures in x. Use the same rule for ln x. For example, $\log_{10} 4.23 \times 10^{-3} = -2.374$. (The mantissa is the '.374' part.)

> **Expert tip**
>
> The mean cannot be more precise than the original measurements. For example, the maximum number of significant figures in the mean when averaging measurements with 3 sf is 3.

> **Expert tip**
>
> If you multiply 4.136 cm, which has four significant figures, by π, you should use 3.1416 which has 5 significant figures for π and your answer, 13.56 cm, will have 4 significant figures.

> ### Worked example
>
> Calculate the pH of $0.052\,mol\,dm^{-3}$ nitric(v) acid. Give your answer to an appropriate number of significant figures.
>
> $pH = -\log_{10}[H^+(aq)] = -\log_{10}(0.052) = 1.283\,996\,656$ (calculator answer)
>
> The number of significant figures in $0.052\,mol\,dm^{-3}$ is 2, so according to the above rule, the mantissa of the logarithm must also contain 2 significant figures. Hence pH = 1.28.

Relating this to significant figures in a pH calculation, if $[H_3O^+(aq)] = 2.22 \times 10^{-3}\,mol\,dm^{-3}$ (that is, where there are 3 significant figures in the mantissa: 2.22), then pH = $-\log_{10}(2.22 \times 10^{-3}) = 2.654$ (3 significant figures are replicated after the decimal point).

The number of significant figures in 10^x is the number of significant figures in the mantissa of x.

> ### Worked example
>
> The equilibrium constant for a reaction at two different temperatures is 0.0322 at 298.2 K and 0.473 at 353.2 K. Calculate the value of $\ln\left(\dfrac{k_2}{k_1}\right)$.
>
> Both rate constants have 2 significant figures, so the ratio $\dfrac{k_2}{k_1}$ should also have 2 significant figures: $\dfrac{k_2}{k_1} = 13.89$. So $\ln\left(\dfrac{k_2}{k_1}\right)$ should have 2 significant figures in the mantissa. $\ln\left(\dfrac{k_2}{k_1}\right) = \ln 13.89 = 2.63$.

Errors

■ Accuracy and precision

Figure 4.2 illustrates the difference between accuracy and precision using a dartboard analogy: the centre represents the true value. Lack of precision causes random errors and lack of accuracy causes systematic errors.

> ### Key definitions
>
> **Accuracy** – measurements or results that are accurate are in close agreement with the true or accepted values.
> **Precision** – precise measurements are ones in which there is very little spread about the mean value.

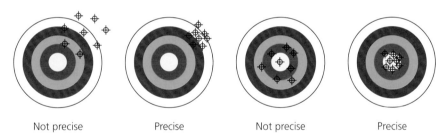

| Not precise | Precise | Not precise | Precise |

Figure 4.2 Accuracy versus precision

■ Uncertainties

■ Sources of uncertainty

Random errors usually arise from the limit of accuracy of the apparatus or instrument. They arise from fluctuations that (on average) cause about half the measurements to be too high and about half to be too low.

Generally, you should report the uncertainty or random error associated with a reading from a scale as ± the smallest division.

For example, a temperature measured with a thermometer has an uncertainty of $\pm\,0.5\,°C$ if the graduations (marks) are $1\,°C$ apart. It may record a temperature of $(24.0 \pm 0.5)\,°C$.

For a digital reading, the last digit is taken as the random uncertainty.

For a digital meter, the random uncertainty is the smallest reading that the meter can display. The uncertainty is quoted to the same number of decimal places as the value (for example, $2.40 \pm 0.01\,V$).

The random uncertainties for the thermometer and voltmeter readings are **absolute errors** and have units. Raw data should be recorded with absolute random uncertainties.

However, you may find that the manufacturer's random uncertainty (also known as the tolerance) printed on the glassware or instrument does not match up with your predicted random uncertainty from these two simple rules. You could use either in your error **propagation**, but you must state the origin of your random errors in your report.

■ Reducing random errors

To minimize the effect of random errors, each measurement can be repeated a number of times (known as a trial) and the results averaged. Choosing a more precise apparatus or instrument would also minimize this random error (Table 4.1).

Random error	Effect	How to minimize
Fluctuations on an electronic balance due to air currents moving over the weighing pan	Readings will be higher and lower than the actual value.	Close the draft shield and make sure the reading is stabilized before making a recording. Repeat the measurement a number of times. Analyze the averages.
Inexperience in reading a scale	Readings will be higher and lower than the actual value.	Repeat the measurement a number of times to enable more accurate readings. Analyze the averages.
Estimating between an increment on an analogue scale, such as a burette or measuring cylinder	Inexperience will cause the readings to be higher and lower than the actual value.	Repeat the measurement a number of times and analyze the averages. Only attempt to estimate to half an increment on the scale.
Volume uncertainties on a measuring cylinder	The size of the uncertainty will depend on the quality of the measuring cylinder and cause the readings to be higher and lower than expected.	Use an 'A' grade measuring cylinder instead of 'B' grade. Choose a measuring cylinder with a lower random uncertainty or use a micropipette. Repeat the measurement and analyze the averages.
Difficulty in reading the scale on a ruler	Readings will be higher and lower than the actual value.	Take repeated measurements and analyze the averages. Select a measuring instrument with greater precision, such as Vernier calipers.
Rounding in a calculation	Readings will be higher or lower depending on whether you round up or down.	Leave one extra significant figure in a calculation and then round to a number of significant figures that is consistent with the input measurements when expressing the final result.
Uncertainties in temperature on a thermometer	Reported as half an increment on the scale, typically $0.5\,°C$. The random uncertainty will place the temperature measurement above and below what is measured.	Use a more precise temperature probe which has a lower random uncertainty.
Recording the time with a stopwatch	Human reaction times are much slower than a stopwatch, which can measure to the nearest $0.01\,s$. Therefore the measured times will be above and below the actual time.	Repeat the measurement a number of times and analyze the averages. An estimate of ±0.3 seconds could be used as the approximate uncertainty.
Visual perception or subjective colour change as the end-point of the reaction, for example, precipitation, is not abrupt (difficult to judge colour change)	Time taken to measure the extent of the precipitation reaction may be higher or lower than the actual value.	Use a colorimeter or spectrophotometer / use a control for colour / repeat experiments.
Inconsistent rate of stirring	Faster stirring increases the chances of collisions between reacting particles and thus causes the time taken to be shorter and vice versa.	Use a magnetic stirrer to ensure consistent stirring of the reactants.

Table 4.1 Minimizing random errors

Examiner guidance

There are two grades of glassware: A and B, with different random uncertainties (tolerances). For example, a class A 25 cm³ pipette has an error of ±0.05 cm³, but a class B one has an error of ±0.10 cm³.

The **resolution** of an instrument, such as a manually operated electronic stopwatch, may not be the limiting factor in the random uncertainty in a measurement.

Worked example

A stopwatch has a resolution of hundredths of a second, but the random uncertainty is due to the human reaction time (≈ ±0.3 s). You should write the full reading (for example, 13.10 s), carry the significant figures through for all repeats, and reduce to an appropriate number of significant figures after an averaging process later (for example, 13.1 s or, more realistically, 13 s).

Expert tip

Sometimes the uncertainty must be predicted at a reasonable level depending on the circumstances. For example, if you are using a pipette (±0.05 cm³) to measure the volume of a solution containing escaping carbon dioxide gas, the uncertainty may increase to ±0.2 cm³.

■ ACTIVITY

13 The temperature of a mixture of crushed ice, sodium chloride and water is measured using a mercury-in-glass thermometer (Figure 4.3). Deduce the temperature of the mixture and the associated random uncertainties from the scale (with and without interpolation). Find out about the relative disadvantages of using alcohol versus the relative advantages of using mercury in liquid-in-glass thermometers.

Figure 4.3 Measurement of a freezing mixture using a mercury thermometer

Examiner guidance

You must recognize that all measured values have uncertainty and are not exact.

■ Errors not due to errors in measurement

One example is in detecting the end-point of a titration by judging when the indicator just changes colour. If you estimate the end-point in a titration to within one drop of the titrant (≈ 0.05 cm³) then the absolute random uncertainty will be approximately ±0.05 cm³. The end-points of some titrations, for example, EDTA titrations, are difficult to detect and you may be justified in using ±0.10 cm³.

Common mistake

A common mistake is to regard measurements as pure numbers with exact values. They are physical quantities with units and an uncertainty (error).

Molar masses are experimentally determined quantities and it can be assumed that the uncertainty in the molar mass of a substance is ±1 in the last significant digit. Consider potassium chloride, KCl. If relative atomic masses quoted to 2 decimal places (as in the IB chemistry data booklet) are used, then the molar mass of potassium chloride is calculated as 74.55 g mol^{-1}. This corresponds to a random percentage uncertainty of $\left(\dfrac{0.01}{74.55}\right) \times 100 = 0.013\,\%$. However, this is usually insignificant compared to the random uncertainty in the mass reading and hence can be ignored (or briefly mentioned).

■ Percentage error

Chemists establish the accuracy of their measurements by comparing their results with values that are well established in the chemical literature and are considered to be 'accepted values'.

Taking x to be the experimentally measured value and y to be the accepted value:

$$\text{Percentage error} = \left| \frac{x - y}{y} \right|$$

■ ACTIVITY

14 A redox titration of brass nails has an experimentally determined percentage by mass of copper of 39.7 %. The manufacturer claims that the brass nails contain 44.2 % copper by mass. Calculate the percentage error in the result.

■ Combining random uncertainties (error propagation)

Random uncertainties should be combined using the rules outlined in Table 4.2.

Combination	Operation	Example
Add or subtract values	Add the absolute uncertainties	Mass of weighing bottle and silver nitrate = (18.54 ± 0.01) g
		Mass of weighing bottle = (13.32 ± 0.01) g
		Mass of silver nitrate = (5.22 ± 0.02) g
Multiply values	Add the percentage uncertainties	Mass of water = (50.0 ± 0.1) g
		Temperature rise = (10.8 ± 0.1) °C
		Percentage random uncertainty in mass = 0.20 %
		Percentage random uncertainty in temperature = 0.93 %
		Enthalpy change = 2259 J
		Percentage random uncertainty in heat change = 1.13 %
		Absolute random uncertainty in heat change = ±26 J
		(Note: the random uncertainty in the specific heat capacity of water (4.184 J g^{-1} °C^{-1}) is taken to be zero.)
Divide values	Add the percentage uncertainties	Mass of sodium chloride in solution = (100.0 ± 0.1) g
		Volume of sodium chloride solution = (250.0 ± 0.5) cm^3
		Percentage random uncertainty in mass of sodium chloride = 0.10 %
		Percentage random uncertainty in volume of solution = 0.20 %
		Concentration of sodium chloride solution = 0.400 g cm^{-3} = 40.0 g dm^{-3}
		Percentage random uncertainty of concentration of sodium chloride = 0.30 %
		Absolute random uncertainty of concentration of sodium chloride = ±0.0012 g cm^{-3} = 1.2 g dm^{-3}
Power rules	Multiply the percentage uncertainty by the power	[H$^+$(aq)] = 0.152 ± 0.001 mol dm^{-3}
		Rate of reaction = k[H$^+$(aq)]^2 = 0.213 mol dm^{-3} s^{-1}
		(Note: the uncertainty in the rate constant, k, is taken as zero and its value in this reaction is 0.922.)
		Percentage uncertainty in concentration = 0.66 %
		Percentage uncertainty in rate = 1.32 %
		Absolute uncertainty in rate = ±0.003 mol dm^{-3} s^{-1}

Table 4.2 Propagation of random errors

- **Repeated measurements**

Repeating measurements and averaging them to calculate a mean reduces the random uncertainty. The final answer could be given to the propagated error of the component values in the average.

Worked example

ΔH_{mean}

$$= \frac{[+100\,kJ\,mol^{-1}\,(\pm10\,\%) + 110\,kJ\,mol^{-1}\,(\pm10\,\%) + 108\,kJ\,mol^{-1}\,(\pm10\,\%)]}{3}$$

$= 106\,kJ\,mol^{-1}\,(\pm10\,\%)$

- **Multiple readings**

If more than one reading of a measurement is made, then the random uncertainty increases with each reading. For example, $10.00\,cm^3$ of acid is delivered from a $10\,cm^3$ pipette $(\pm0.10\,cm^3)$, repeated 3 times. The total volume delivered is $30.00 \pm 0.30\,cm^3$.

- **Fluctuating readings**

The measurements on an electronic balance will fluctuate. Start with the numbers that are not fluctuating and then make a guess as to what the next digit would be. For example, if a balance gives the following readings: $12.345\,g$, $12.320\,g$, $12.349\,g$, $12.357\,g$ and $12.327\,g$, the average could be reported as $12.34 \pm 0.05\,g$.

■ ACTIVITY

A piece of steel, $2.923\,g \pm 0.002\,g$, was reacted and dissolved in $50.00\,cm^3 \pm 0.10\,cm^3$ of sulfuric acid $(2.00\,mol\,dm^{-3} \pm 0.02\,mol\,dm^{-3})$; the sulfuric acid is in excess. The resulting mixture was filtered and the filtrate was diluted to $250.00 \pm 0.30\,cm^3$ using a volumetric flask.

$Fe + H_2SO_4 \rightarrow FeSO_4 + H_2$

$MnO_4^- + 5Fe^{2+} + 8H^+ \rightarrow Mn^{2+} + 5Fe^{3+} + 4H_2O$

$20.00\,cm^3 \pm 0.03\,cm^3$ of the diluted solution was titrated with $0.0500\,mol\,dm^{-3} \pm 0.0005\,mol\,dm^{-3}$ acidified potassium manganate(vii) solution. Four titrations were carried out and the data collected are shown below.

Trial	1	2	3	4
Final burette reading/cm $\pm 0.05\,cm^3$	15.50	35.20	15.30	35.20
Initial burette reading/cm $\pm 0.05\,cm^3$	0.00	20.00	0.00	20.00
Volume used of manganate(vii) ions/cm $\pm 0.10\,cm^3$	15.50	15.20	15.30	15.20

15 Calculate the percentage by mass of iron in the steel and perform error propagation.

RESOURCES

You can use this website to make calculations involving uncertainties:

http://web.mst.edu/~gbert/JAVA/uncertainty.HTML

■ ACTIVITIES

An empty crucible was weighed, using an electronic balance with a random error of ±0.01 g for a single reading. The crucible was reweighed containing some pure solid metal carbonate, MCO_3, and then heated to constant mass. The metal carbonate undergoes thermal decomposition and decomposes according to the equation:

$$MCO_3(s) \rightarrow MO(s) + CO_2(g)$$

The following raw numerical data was obtained (by weighing by difference):

Mass of empty crucible = 15.23 g ± 0.02 g

Mass of empty crucible and metal carbonate = 17.46 g ± 0.02 g

Mass of crucible and contents after heating to constant mass = 16.61 g ± 0.02 g

16 Calculate the mass (in grams) of carbon dioxide released.

17 Hence, calculate the amount in moles of carbon dioxide produced.

18 Calculate the mass of metal carbonate taken.

19 Calculate the molar mass of the metal carbonate and hence the molar mass of the metal, M.

20 Assuming all the single measurements of mass (including 'zero') had a random uncertainty of ±0.01 g calculate the random uncertainty in the experimental value of the molar mass of M.

■ Error propagation in logarithmic and trigonometric functions

One simple approach is to take the error as the greatest deviation.

For example, if $S = x \cos \theta$ with $x = (3.0 \pm 0.2)$ cm and $\theta = (47 \pm 2)°$

$S = 3.0 \cos 47° = 2.046$

$S_{max} = 3.2 \cos 45° = 2.263$; $S_{min} = 2.8 \cos 49° = 1.837$

$S - S_{max} = 2.046 - 2.263 = -0.217$; $S - S_{min} = 2.046 - 1.837 = 0.209$

Therefore, $S = (2.0 \pm 0.2)$ cm

■ Systematic errors

A systematic error causes **all** of the measurements to be higher or **all** of them to be lower than the true result (a bias). Systematic errors can arise from:

■ **Flaws in the equipment or apparatus used in recording measurements**

If a stopwatch used to record times for chemical reactions (for example, sulfur obscuring a cross) is running fast, this means that the time for the reaction (the time to reach a certain point in the reaction) will be underestimated since all the readings will be less than the true value.

■ **Flaws in the procedures used**

If a 25 cm³ pipette is used at a temperature of 55 °C rather than at the calibration temperature of 20 °C, then the volume it delivers will be consistently above 25.00 cm³ as a result of its thermal expansion (although it should be noted that this does not represent a major source of error in most situations).

■ **Lack of purity in reagents**

Potassium manganate(VII) is not a primary standard and its concentration decreases with time (due to hydrolysis and photodecomposition). If the potassium manganate(VII) has not been re-standardized just before a titration then this will cause a systematic error.

Systematic errors can only be minimized by changing the way the experiment is carried out (Table 4.3). They cannot be minimized by recording more measurements and averaging. A systematic error is often difficult to detect but once it has been detected, it can usually be corrected and thus eliminated.

> **Expert tip**
>
> **Measurement error** includes random and any systematic errors.

Systematic error	Effect	How to minimize
Using an electronic balance that does not display zero grams. For example, the zero reading is 0.010 g (does not tare correctly). This is a **zero error**.	The reading will be higher or lower than the real value.	Fix (calibrate) the balance so that it can be zeroed.
Parallax error: incorrectly reading the meniscus on a measuring cylinder (Figure 4.4) or burette.	The reading will be higher or lower than expected, depending on how it is being viewed.	Make sure you view the meniscus at eye level.
Using an old wooden ruler for measuring distance.	The reading will always deviate by the same amount. Depending on the ruler, the readings will either be higher or lower than expected. This is another example of a zero error.	Use Vernier callipers because these are more precise with a greater sensitivity.
Heat is lost to the environment from the chemicals reacting in a solution.	The temperature of the reaction will always be lower than expected.	Insulate the reaction flask with polystyrene.
Overshooting the end-point in a titration.	The reading will always be higher than expected.	Making sure that the titration is stopped at the same colour change each time. Use a pH probe if appropriate.
Having to make approximations in measurements.	The reading might be higher or lower than expected.	Make the same approximation each time.
Faulty apparatus (for example, a thermometer with an air bubble in it, or a blocked gas delivery tube).	The reading will be higher or lower than expected.	Make sure the equipment is fixed so it is working correctly.

Table 4.3 Identification and minimization of common systematic errors

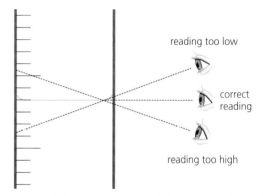

reading too low

correct reading

reading too high

Figure 4.4 Parallax error with a measuring cylinder

■ ACTIVITIES

21 In a series of acid–base titrations, a student always used 15 drops of indicator per titration (rather than the recommended 2 drops). State and explain whether this would result in a random or systematic error in the titre volumes.

22 A student calibrated a pH meter with an acidic buffer labelled as 4.00 and then measured the pH values of various aqueous solutions. The student later discovered that the pH of the buffer was 4.10. State and explain whether this would result in a random or systematic error in the pH measurements.

■ ACTIVITIES

A 0.500 g sample of a weak and non-volatile acid, HA, was dissolved in water to make 50.00 cm³ of solution. The solution was then titrated with a standard NaOH solution. Predict how the calculated molar mass of HA would be affected (too high, too low, or not affected) by the following laboratory procedures. Explain each of your answers.

23 After rinsing the burette with distilled water, the burette is filled with the standard NaOH solution; the weak acid HA is titrated to its equivalence point.

24 Extra water is added to the 0.500 g sample of HA.

25 An indicator that changes colour at pH 5 is used to show the equivalence point.

▨ Graphing

Line graphs show relationships and can be used to determine the intercept and gradient (rate of change) (see Chapter 1 for the determination of gradients on a nonlinear graph). The independent variable is plotted along the *x*-axis and the dependent variable along the *y*-axis. Graphs may be generated using Excel® or data-logging software (Chapter 5).

Axes should be labelled with the quantity being measured and the units, separated with /.

Data points (>5) should be marked with a cross (× or +) and error bars.

Spread the data points on a graph as far as possible without using scales that are difficult to deal with (for example, multiples of 3, 7 and 11).

Consider the maximum and minimum values of each variable, the size of the graph, and whether the origin should be included as a data point.

Figure 4.5 shows some right and wrong ways to plot lines and curves of best fit that reduce the effect of random errors. A line of best fit must pass through the origin for a directly proportional relationship.

■ Random uncertainties from gradients in linear graphs

To determine the uncertainty in the gradient of a linear graph (with error bars), two gradients should be drawn on the line graph as shown in Figure 4.6. If possible, the line of best fit should be drawn by Excel® rather than by hand. Then add the steepest (shown in blue) or shallowest (shown in red) gradient line that can be drawn through the data points (with error bars).

Percentage uncertainty in gradient

$$= \frac{\text{highest (or lowest) gradient} - \text{gradient of line of best fit}}{\text{gradient of line of best fit}} \times 100$$

Examiner guidance

Interpolation is an estimate of data point values within the existing range of data. Extrapolation is estimating data points from beyond the range of your data set. Figure 4.7 distinguishes between interpolation and extrapolation, assuming the stated relationship between variables holds beyond the experimental range of results. It is more accurate when done with a line of best fit than with a curve of best fit.

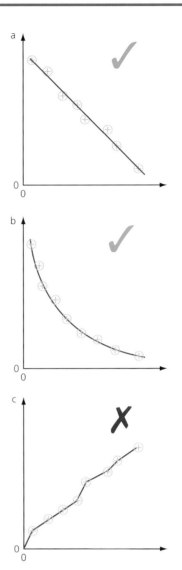

Figure 4.5 Right and wrong ways to draw lines of best fit

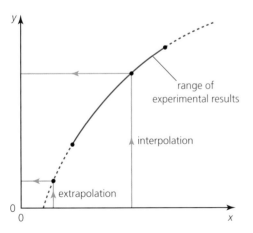

Figure 4.7 Interpolating and extrapolating

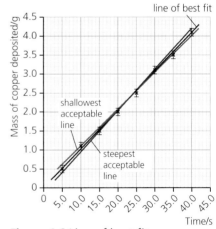

Figure 4.6 Line of best fit, shallowest acceptable gradient and steepest acceptable gradient for a graph from the electroplating of copper on an inert electrode

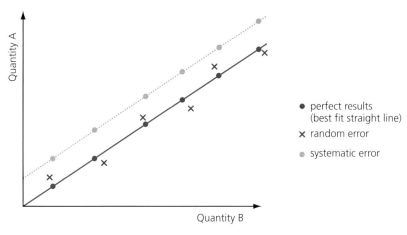

Figure 4.8 Perfect results (no errors), random uncertainties and systematic errors (positive bias) of two directly proportional quantities

Expert tip

You can also use straight line graphs to identify systematic errors when data for an expected linear and directly proportional relationship does not pass through the origin (Figure 4.8).

■ Linearizing raw data

A chemical relationship may be linear (proportional or directly proportional) or some type of curved graph, such as an inverse relationship (Figure 4.9), or a logarithmic (Figure 4.10) or exponential relationship (Figure 4.11). Often the data for a graph with a non-linear relationship is transformed to a linear plot.

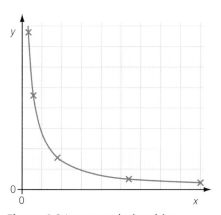

Figure 4.9 Inverse relationship ($y = 1/x$)

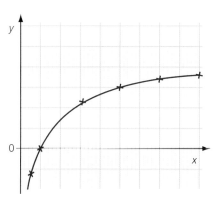

Figure 4.10 Natural logarithm relationship ($y = \ln x$)

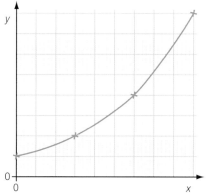

Figure 4.11 Exponential relationship ($y = e^x$)

It is usually possible to manipulate the mathematical form of data such that an easily analyzed straight line results when it is plotted (Table 4.4). Essentially the approach is to obtain a relationship in the form $y = mx + c$.

Example of non-linear function	Linear form and linear plot ($y = mx + c$)
$y = ax^b$	$\ln y = \ln a + b \ln x$
	$\ln y$ versus $\ln x$; gradient = b; intercept = $\ln a$
$y = ab^x$	$\ln y = \ln a + x \ln b$
	$\ln y$ versus x; gradient = $\ln b$; intercept = $\ln a$
$y = x^{-1}$	$y = x^{-1}$
	y versus $1/x$; gradient = 1; intercept = 0

Table 4.4 Using logs to obtain a linear form for a non-linear relationship

Figure 4.12 shows averaged raw data for the reaction between sodium thiosulfate and acid, warmed and cooled together at different temperatures of sodium thiosulfate.

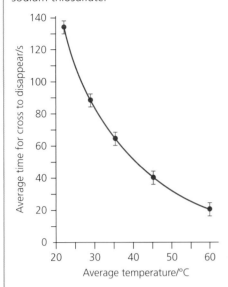

Figure 4.12 A graph of average time for the cross to disappear, against average temperature for the reaction between sodium thiosulfate and acid

The dependent variable, time, is converted into a processed variable, average rate, which is the reciprocal of the time. A linear and directly proportional relationship (Figure 4.13) is obtained and allows the accuracy of the data to be easily observed or calculated.

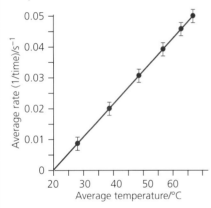

Figure 4.13 A graph of average rate (time^{-1}) against average temperature for the reaction between sodium thiosulfate and acid

◼ Rejection of data

You may find during an investigation that one result in a set of measurements does not agree well with the other results. You must decide how large the difference between the suspect result and the other data must be before you discard the result as an outlier (Figure 4.14). One simple approach is known as the two standard deviation test.

Calculate the mean (average) and standard deviation for your data. Any data value equal to or greater than two standard deviations from the mean value may be rejected with a high percentage of confidence.

◼ ACTIVITY

26 Consider the following set of measurements: 0.1012, 0.1014, 0.1012, 0.1021, 0.1016.

Apply the two standard deviation test to decide whether 0.1021 should be discarded.

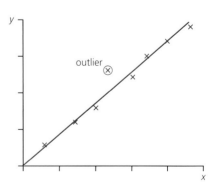

Figure 4.14 A possible outlier on a linear graph

5 Information communication technology

Spreadsheets

■ Functions in Excel®

There are a number of useful mathematical and statistical functions in Excel® that can be used to analyze raw data and construct mathematical models. These can carry out specific calculations (for example =B6*B7/SQRT(B4) multiplies the contents of cells B6 and B7 and divides the result by the square root of cell B4). Equations and functions in spreadsheets are dynamic: the results change if the source data is changed. Table 5.1 lists some other useful built-in Excel® functions.

=SUM(A2:A5)	Find the sum of values (numbers) in the range of cells A2 to A5
=COUNT(A2:A5)	Count the number of numbers
=COUNTIF(A1:A10,100)	Count cells equal to 100
=COUNTIF(A1:A10,'>30')	Count cells greater than 30
=AVERAGE(A2:A5)	Find the mean (average) of the numbers
=AVEDEV(A2:A5)	Find the average deviation from the mean of the numbers
=STDEV(A2:A5)	Find the sample **standard deviation** of the numbers
=VAR(A2:A5)	Find the variance of the numbers
=MAX(A2:A5)−MIN(A2:A5)	Find the **range** of the numbers
=LN(A1)	Find the natural logarithm of the number in cell A1
=LOG(A1)	Find the logarithm (to the base 10)
=EXP(A1)	Returns *e* raised to the power of the number

Table 5.1 Useful built-in Excel® functions

■ Graphing in Excel®

Select the block of cells in Excel® containing the data to be plotted (which may include headings). The *x*-axis data column should always be to the left of the *y*-axis. Click on the **Insert** tab and choose the graph type from the **Charts** area, usually **Scatter** if the data are continuous. Graphs created on a separate Excel® sheet can be copied or pasted into a Word® document. Graphs embedded into an Excel® worksheet can be edited even after they have been inserted.

Experimental data will show scatter due to random errors in the measurement. A trend line can be added by right-clicking on a data point and then selecting **Add Trendline**. If the data lies on an approximately straight line and a linear relationship is expected, then select **Linear** for the **regression type**. Check the relevant boxes to display the equation of the line (**regression** equation) and the correlation value (R^2).

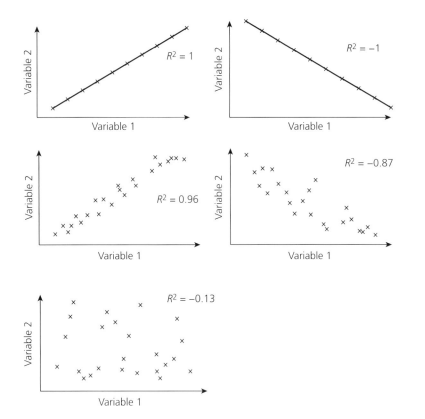

Figure 5.1 The strength of the correlation in the linear scatter graphs is the correlation coefficient that extends from +1 to −1

■ **ACTIVITY**

Type in potassium chromate(VI)/mol/dm³ in the A1 cell (*x*-axis) and absorbance in the B1 cell (*y*-axis). Enter the data for concentration (from A2 to A6) and absorbance (B2 to B6). Use Format > Cells > Number to adjust the number of decimal places to 3.

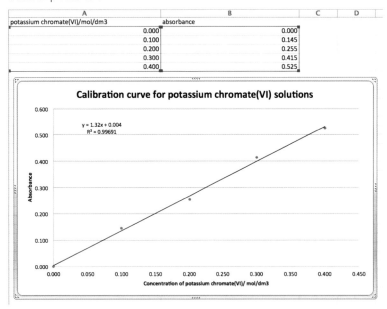

Figure 5.2 Screenshot of the process

Highlight the concentration and absorbance columns and click on Insert on the top menu; then click on Chart on the drop-down menu. Click on the *xy* scatter option and click Finish. Right click in the graph area and click on Chart Options. You can remove the legend and add gridlines, a graph title, and labels for the *x* and *y* axes.

Click on Chart in the top menu and click on the Add Trendline option. This will put in the linear regression line or best-fit straight line through the data points. Click on Automatic, Display equation on chart, and Display *R*-squared value on chart. The screenshot in Figure 5.2 shows the data and graph created.

■ Modelling

Excel® can be used to model chemical phenomena, such as statistical entropy. The spreadsheet in Figure 5.3 calculates the number of microstates available to a mixture of particles A and B at equilibrium. Particles of A cannot be distinguished from each other and neither can particles of B. ln(W) is proportional to the entropy (S).

The number of microstates (W) for each combination of particles

$$= (A + B)!/(A! \times B!)$$

Excel® uses =FACT(N) to calculate N factorial, for example:

$$5! = 5 \times 4 \times 3 \times 2 \times 1$$

A Particles of A	B Particles of B	C (A+B)!		D (A! * B!)		E W= (A+B)!/(A! * B!)	F LN(W)	G
20	0		2.4329E+18		2.4329E+18	1	0	
19	1		2.4329E+18		1.21645E+17	20	2.995732	
18	2		2.4329E+18		1.28047E+16	190	5.247024	
17	3		2.4329E+18		2.13412E+15	1140	7.038784	
16	4		2.4329E+18		5.02147E+14	4845	8.485703	
15	5		2.4329E+18		1.56921E+14	15504	9.648853	
14	6		2.4329E+18		6.27684E+13	38760	10.56514	
13	7		2.4329E+18		3.13842E+13	77520	11.25829	
12	8		2.4329E+18		1.93133E+13	125970	11.7438	
11	9		2.4329E+18		1.4485E+13	167960	12.03148	
10	10		2.4329E+18		1.31682E+13	184756	12.12679	
9	11		2.4329E+18		1.4485E+13	167960	12.03148	
8	12		2.4329E+18		1.93133E+13	125970	11.7438	
7	13		2.4329E+18		3.13842E+13	77520	11.25829	
6	14		2.4329E+18		6.27684E+13	38760	10.56514	
5	15		2.4329E+18		1.56921E+14	15504	9.648853	
4	16		2.4329E+18		5.02147E+14	4845	8.485703	
3	17		2.4329E+18		2.13412E+15	1140	7.038784	
2	18		2.4329E+18		1.28047E+16	190	5.247024	
1	19		2.4329E+18		1.21645E+17	20	2.995732	
0	20		2.4329E+18		2.4329E+18	1	0	

Figure 5.3 Simulation of statistical entropy

Excel® can also be used to model a range of dynamic chemical phenomena, such as kinetics and equilibrium. The spreadsheet is an excellent tool when a large number of repetitive calculations need to be made (such as variables that are changed incrementally), and data must be presented graphically. In chemical kinetics and equilibrium, the numerical solution of the rate laws can easily be implemented using iterative equations of the type:

new value = old value + rate of change × small time interval

However, spreadsheets have their limitations and you may want to use a programming language such as Python.

Python programming

More than 50 percent of jobs require some degree of technology skills, and this percentage looks set to increase in the coming years, particularly in STEM (science, technology, engineering, mathematics) and related fields. This is a good reason to gain some experience with basic coding skills while studying IB Diploma group 4 subjects.

Many topics in chemistry lend themselves to coding-type activities, including modelling the shapes of molecules in 3D, creating visualizations and graphs of periodic trends (for example, atomic and ionic radii), or modelling a mathematical relationship such as the Nernst equation or a titration curve. Indeed, modelling a 'theoretical curve' for comparison with experimentally derived data could serve as a useful adjunct for many different IA investigations.

One language that lends itself to this kind of visual modelling is Python, or more specifically, VPython, which is the Python programming language plus a 3D graphics module called Visual. There are a number of online IDEs (integrated development environments) where students can code using Python and/or VPython without the requirement to download any software. These include Cloud9 (c9.io), GlowScript (glowscript.org) and Trinket (trinket.io).

As an indication of the power of VPython, the six lines of code in Figure 5.4 were used to create a 3D model of a methane molecule using the Trinket IDE (Figure 5.5). After running the code, the model can be rotated in three dimensions by right clicking on it.

⟨ ⟩   main.py

```
1  GlowScript 2.6 VPython
2
3  scene.background = color.yellow
4  carbon = sphere(color = color.black, pos = vector(0,0,0), radius = 1)
5  H1 = sphere(color = color.white, pos = vector(0.629118,0.629118,0.629118), radius = 0.75)
6  H2 = sphere(color = color.white, pos = vector(-0.629118,-0.629118,0.629118), radius = 0.75)
7  H3 = sphere(color = color.white, pos = vector(0.629118,-0.629118,-0.629118), radius = 0.75)
8  H4 = sphere(color = color.white, pos = vector(-0.629118,0.629118,-0.629118), radius = 0.75)
9
```

Figure 5.4 Code for creating a 3D model of a methane molecule in VPython using the Trinket IDE

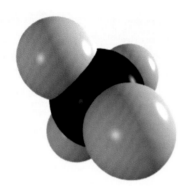

Figure 5.5 3D model of a methane molecule created by the code in Figure 5.4

The code in Figure 5.6 is used to model the titration curve for a strong acid–strong base titration.

```
1  GlowScript 2.6 VPython
2
3  g1 = graph(title = "pH curve for the reaction of 0.1M HCl with 0.1M NaOH",xtitle="vol. of NaOH [cm\u00B3]",ytitle="pH")
4  f1 = series(color=color.blue) #alternatively change 'gdots' to 'series' for a "joined up" line graph
5  c_HCl_init = 0.1 #initial concentration of HCl in mol/dm^3
6  c_NaOH_init = 0.1 #concentration of NaOH in mol/dm^3
7  V_NaOH = 0 #volume of NaOH added in dm^3
8  delta_V = 0.1 #volume increment in dm^3
9  V_HCl = 25 #volume of HCl in dm^3
10 V_max = 50 #maximum volume in dm^3
11
12 print('V(NaOH)/cm\u00B3', '     pH') #creates the table headings for the data window
13 #Note: '\u00B3' is the unicode for a supercript 3
14
15 #This section calculates and plots the points before the equivalence point:
16 while V_NaOH < 25:
17     rate(100) #Sets the calculation rate; if this line is removed the graph will be plotted instantly
18     n_NaOH = (V_NaOH/1000) * c_NaOH_init
19     n_HCl_left = ((V_HCl/1000) * c_HCl_init) - n_NaOH
20     c_HCl = n_HCl_left/((V_HCl+V_NaOH)/1000)
21     pH = -log(c_HCl)/log(10)
22     print(V_NaOH,'          ', pH)
23     f1.plot(V_NaOH, pH)
24     V_NaOH += delta_V
25
26 #This section plots the pH at the equivalence point:
27 V_NaOH = 25
28 pH = 7
29 print(V_NaOH, '          ', pH)
30 f1.plot(V_NaOH, pH)
31 V_NaOH += delta_V
32
33 #This section calculates and plots the points after the equivalence point up to the maximum volume:
34 while V_NaOH < V_max:
35     rate(100)
36     n_NaOH = (V_NaOH - V_HCl)/1000 * c_NaOH_init
37     c_NaOH = n_NaOH/((V_HCl+V_NaOH)/1000)
38     pOH = -log(c_NaOH)/log(10)
39     pH = 14 - pOH
40     print(V_NaOH,'          ', pH)
41     f1.plot(V_NaOH, pH)
42     V_NaOH += delta_V
```

Figure 5.6 VPython code used to model a titration curve for a strong acid–strong base titration

The code is divided into four main sections. The first sets the initial values of relevant variables and sets up the graph. The second, third and fourth sections calculate and plot the data before, at and after the equivalence point of the titration. The output of the code is illustrated in Figure 5.7.

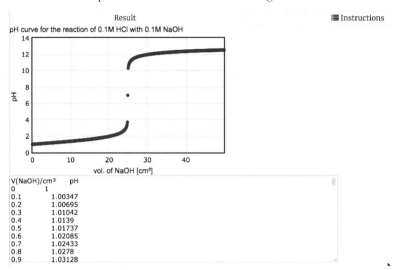

Figure 5.7 Output of the code in Figure 5.6

Note that the data in the window at the bottom of the screen may be copied and pasted into a spreadsheet for further analysis, if so desired. Learning some basic coding skills is a very worthwhile endeavour, and there are some excellent introductory videos on how to use VPython and GlowScript on YouTube, or at www.glowscript.org.

Data-logging

When carrying out experimental work in chemistry, you may need to collect large amounts of raw data over a very long time period or collect raw data over extremely short time intervals. You could use data-logging to investigate titration curves, kinetics including colorimetry, buffers and buffer capacity, and enthalpy changes.

The raw data collected by the probe (sensor) of the data-logger can be stored in a spreadsheet, which can then be used to generate charts and graphs. Most data-logging software will allow for graphs to be generated directly from the raw data.

A sensor measuring a variable (for example, temperature) produces an analogue signal that the computer can convert to digital signals and process digitally. The data-logger acts as an interface between the probe (sensor) and the computer.

Most data-loggers can store data to be retrieved later. This means you can use them remotely from a computer. Data-loggers also incorporate displays that give numerical and graphical information on the readings being made in real-time.

The common probes (sensors) available for use in chemistry include: light level, pH, temperature, dissolved oxygen, gas pressure, oxidation reduction potential (ORP) for redox and potentiometric titration, conductivity, voltage and current.

■ **ACTIVITY**

1 Find out about and describe the technique of a **potentiometric titration**.

Expert tip

Once the raw data has been collected and a graph generated by the software, a process of inquiry should begin. For example, some or all of the following questions may be appropriate:

- For each part of the graph, what was happening during the investigation?
- What caused that peak?
- What are the highest and lowest values?
- How large was a particular change and how long did it take?
- How quickly do the values change?
- What is the underlying trend?
- How does one variable seem to depend on another?
- Are there any obvious outliers or 'noise'?

Databases

A chemical database is a database designed to store chemical information, such as chemical structures, crystal structures from X-ray diffraction data, various spectra, reactions (and mechanisms) and thermophysical data, such as bond enthalpies, melting and boiling points, density, electrical conductivity and enthalpy changes.

Table 26 of the IB Chemistry Data Booklet provides infrared absorption tables that are a good starting point for assigning simple infrared spectra. However, it is often necessary to understand in greater detail some more specific properties of infrared spectra. This may form the basis for an internal assessment centred around databases.

All infrared values are approximate and have a range of possibilities depending on the molecular environment in which the functional group is located. Resonance often modifies a peak's position and peaks are stronger (more intense) when there is a large dipole associated with a vibration in the functional group and weaker in less polar bonds.

The two atoms are one of the major factors influencing the infrared absorption frequency of a bond. The greater the mass of the atoms, the lower the infrared absorption frequency. This effect can be observed in the halogenoalkanes. The spectra of trichloromethane, $CHCl_3$, and deutrotrichloromethane, $CDCl_3$, (Figure 5.8) are another example of this effect. These show isotopic shift.

RESOURCES

● ChemSpider (http://www. chemspider.com) is a free online chemical database maintained by the Royal Society of Chemistry.

● WebSpectra (https://webspectra. chem.ucla.edu/) has an infrared comparison tool that overlays spectra and a search engine to locate specific types of compound based upon IUPAC name, molecular formula or functional group.

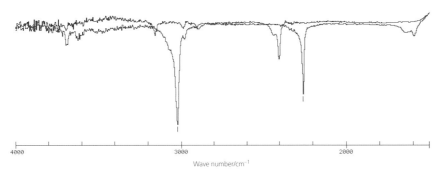

Figure 5.8 The superimposed infrared spectra of trichloromethane (chloroform), $CHCl_3$, and deutrotrichloromethane (deuterochloroform), $CDCl_3$

Two differences between these spectra are the disappearance of the C–H stretching ($3020 \, cm^{-1}$) and C–H bending ($1220 \, cm^{-1}$) in the deuterated compound and a shift relative to the $CHCl_3$. The first is caused by the lack of C–H bonds in $CDCl_3$. The second is because heavier atoms (deuterium (^2_1H) versus protium (^1_1H)) will cause attached bonds to absorb at lower frequencies.

Molecular vibrations can be modelled by visualizing the molecule as consisting of independent pairs of atoms that behave like vibrating point masses connected by springs. The wavenumber of a stretching vibration is given by an equation derived from Hooke's law for a vibrating spring:

$$\bar{v} = 4.12 \sqrt{\frac{k}{\mu}}$$

where k is the force constant ($D \, cm^{-1}$) and μ is the reduced mass of the two

atoms, $\dfrac{(m_1 m_2)}{(m_1 + m_2)}$ where m is the mass of the atoms in grams.

These wavenumbers can be compared to wavenumbers in experimental spectra.

Ideas for investigations involving databases

- Determine whether **cis** and **trans** isomers of alkenes can be identified by bending vibrations.
- Determine whether the absorption of the carbonyl group in cyclic aldehydes shifts to a longer wavelength or higher wavenumber with C–H conjugation.
- Determine whether the effects of ring size and conjugated unsaturation are additive in a range of organic compounds.

Simulations

There is a wide variety of software designed to simulate chemical processes or to illustrate key chemical concepts. Some of these programmes are interactive, so it is possible for you to observe the effects on the simulated system when you change the value of variables.

ChemKinetics (http://pubs.acs.org/doi/pdf/10.1021/ed100373z) is a chemical kinetics simulator and tutorial package that incorporates a range of reversible and irreversible processes with different orders (Figure 5.9).

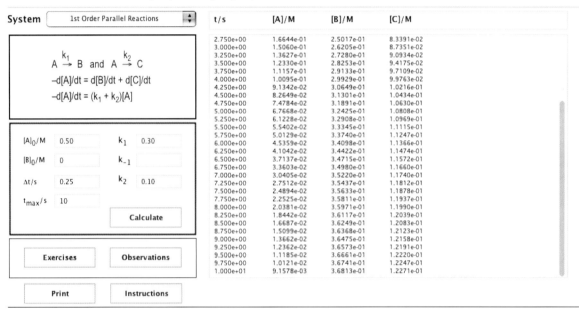

Figure 5.9 Screenshot of ChemKinetics

In a virtual experiment (http://chemcollective.org/activities/autograded/123) the software tabulates data and draws a graph from the experiment (often you can choose apparatus, amounts of chemicals and conditions). You could use this approach to complement your practical work. Alternatively, you could use it as part of your pre-lab discussion to set the scene for your experiment, or to stimulate post-lab evaluation of your experimental process and results.

Excel® can generate dynamic simulations for step-wise equilibria, titrations of acids and bases, concentration plots for polyprotic acids, the Maxwell–Boltzmann distribution curve (for ideal gases), the Nernst equation and the P–V–T surface for an ideal gas (Figure 5.10).

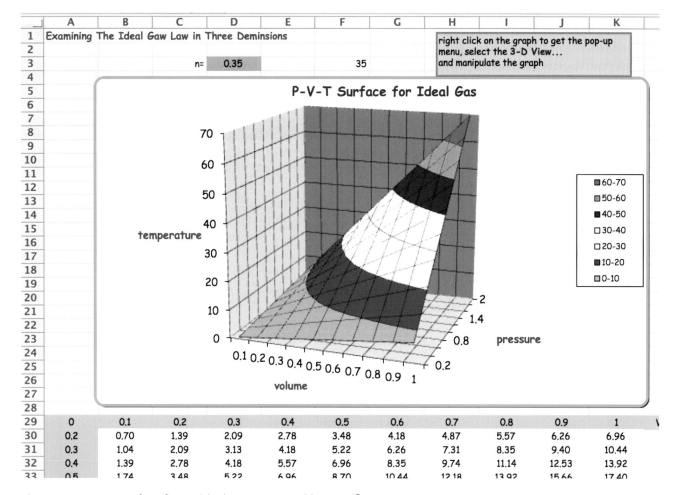

Figure 5.10 *P–V–T* surface for an ideal gas generated by Excel®

Use of smartphones

Your smartphone is a powerful computer and contains an operating system which allows software programs to operate on it ('apps'). There are numerous apps available for smartphones, many with science-specific applications; up to 60% of them are free and many others are available at reasonable prices.

▓ How to obtain apps

Apps can be downloaded from online stores, and online search engines, such as Google, enable you to locate suitable apps, for example, by searching for 'chemistry apps for school/college'. Searching for 'Google Play' brings up Android apps.

Some organizations have developed their own apps, for example:

■ American Association for the Advancement of Science (AAAS): http://sciencenetlinks.com/collections/science-apps

■ National Aeronautics and Space Administration (NASA): www.nasa.gov/connect/apps.html#.U34C6fldV8E

Recommended apps can be found by reading online reviews, for example:

■ www.sellcell.com/blog/five-data-logging-apps-for-schools-and-colleges/

■ www.tomsguide.com/us/pictures-story/962-best-science-apps.html

■ www.wired.com/wiredscience/2008/07/20-iphone-apps

Apps for chemistry

New apps are being developed all the time. Here are some existing apps that you may find useful:

- https://getchemistry.io

- http://thix.co

- https://www.educationalappstore.com/app/category/chemistry-apps

Using smartphones to record data

Sensors are built into smartphones for specific purposes, but specific apps can use the sensors as external measuring instruments for investigations:

- https://itunes.apple.com/us/app/reaction-rate-calculator-for-chemistry-experiments/id909355641?mt=8

- https://itunes.apple.com/us/app/vernier-graphical-analysis/id522996341?ls=1&mt=8

Expert tip

If you use your smartphone to record data for an IA, you need to ensure that the measurements taken are accurate enough to be used in quantitative experiments.

Common mistake

Some apps display graphs that do not have proper axes for the independent and dependent variables and exclude important features such as units.

Writing the internal

Personal engagement
- Selecting an appropriate investigation
- Personal input and initiative

Writing the IA report

Analysis
- Recording a presentation of raw data
- Data processing
- Presenting data – graphing
- Impact of measurement uncertainty
- Interpreting processed data

assessment report

Exploration
- Planning
- Variables
- Background information
- Methodology
- Safety, ethical & environmental issues
- Risk assessments

Evaluation
- Conclusion
- Strengths and weaknesses of the investigation
- Limitations of the data and sources of error
- Improvements and extensions

Communication
- Structure and clarity
- Relevance and conciseness
- Terminology and conventions
- Referencing
- Report format
- Academic honesty

6 Personal engagement

This criterion assesses the extent to which you engage with the exploration and make it your own. Your personal engagement may be recognized in different attributes and skills. These could include addressing personal interests or showing **evidence** of independent thinking, creativity, and demonstrating initiative in the designing, implementation or presentation of your investigation.

Mark	Descriptor
0	The student's report does not reach a standard described by the descriptors below.
1	The evidence of personal engagement with the exploration is limited with little independent thinking, initiative or creativity.
	The justification given for choosing the research question and/or the topic under investigation does not demonstrate personal significance, interest or curiosity.
	There is little evidence of personal input and initiative in the designing, implementation or presentation of the investigation.
2	The evidence of personal engagement with the exploration is clear with significant independent thinking, initiative or creativity.
	The justification given for choosing the research question and/or the topic under investigation demonstrates personal significance, interest or curiosity.
	There is evidence of personal input and initiative in the designing, implementation or presentation of the investigation.

© IBO 2014

Table 6.1 Mark descriptors for the personal engagement criterion

Ideally, for your internal assessment you should design your own individual procedure. An internal assessment can be inspired by an observation, an issue, or a subject area of personal interest. You can demonstrate your personal input and engagement:

- in your research for background information
- by persevering while collecting relevant raw data under difficult circumstances (for example, low sensitivity readings or hard to control variables)
- in your choice of methods of analysis
- when establishing the scientific context of the conclusion (analysis and evaluation).

When considering this criterion you should also bear in mind the following points:

- Your report should have a statement of purpose.
- You need to demonstrate the relationship between your research question and the real world.
- You should show originality in the design of your methodology (choice of materials, instruments and method and their justification).

This criterion is marked using a holistic approach, based on the contents of the full report for the individual investigation. It will therefore overlap with components of other criteria, for example:

- **Exploration**: in your selection and application of analytical and graphical techniques to process your numerical data
- **Analysis**: by your comments concerning the quality of your raw data, and the type of material you refer to in the background or in the discussion of the results
- **Evaluation**: through the depth of understanding you show in assessing the limitations of your investigation, and through your reflective comments on the improvements and extensions to your investigation.

Common mistake

Avoid contriving the personal significance of your investigation without giving further concrete reasons for the choice (for example, it is not sufficient to simply state 'I have always been interested in...'; you should demonstrate your personal connections to the topic.

Expert tip

You cannot use a classic investigation (well-known and published method) unless you attempt to modify it. The subject for the investigation must be **original** in some way. Personal input can be reflected at the simplest level by having completed the investigation, but you cannot expect to score highly if you simply follow a classic experiment with no sign of application or if you are unable to troubleshoot it. There must be some indication in your report that you showed commitment to your investigation.

The following guiding questions may help you develop a plan for your individual investigation and incorporate personal engagement:

- What chemical phenomenon, chemical reaction, chemical process or system are you going to investigate?

- Why is it worthwhile or justified to investigate this?

- Why are you personally interested?

- Are there opportunities to show personal engagement skills, such as independent research and thinking?

- What mathematical, graphical and simulation skills do you need to apply?

- What is already known in the chemical literature?

- What new chemical knowledge is currently being investigated in this area?

- Is your research question answerable within the constraints of time and resources in your chemistry department?

- Which method will you use and adapt: will you apply an 'established' methodology to a new topic or apply a 'new' methodology to an established topic?

- Can the investigation be organized into a sequence of experiments?

- How many samples will you be able to analyze?

- How many experiments will you conduct? How long will each experiment take?

- Does an overall time of 10 hours give you enough time to complete all the experiments in your lessons?

- How will you record and organize your raw data and observations?

- Will you use a data-logger and probes to record some raw data?

- What apparatus, instruments and techniques will you use?

- What chemicals and solvents do you need? Are they safe to use (with suitable safety precautions) and stable? How should the chemicals be handled and disposed of after practical work?

- Do you know how to operate the apparatus, such as a rotatory evaporator, colorimeter or UV-vis spectrophotometer?

- How will you control and monitor the controlled variables?

- Can you find secondary data?

- Can you simulate some aspects?

■ Justifying your research question

Your topic of study may be of significance to a community and an economy (for example, biofuels and the synthesis of pharmaceuticals), or to the environment (for example, disposal and recycling of batteries). Your research question could be based upon an observation (for example, browning of bananas when exposed to air, which is connected to ripening, transport of fruit and the hormone, ethene), or an interesting demonstration, such as an autocatalytic clock reaction (connected to reaction–diffusion systems of chemicals in developing embryos).

■ Evidence of personal input

You need to show evidence of independent thinking, personal input and initiative in the design of your investigation and/or in its implementation. You can show this through the level of commitment you show throughout the whole process, including your persistence in data collection, design of apparatus, data analysis or modification of techniques.

Expert tip

Do not forget that the total time allocation for your internal assessment is 10 hours: this includes planning time, implementation, and writing your report.

RESOURCES

Ideas for practical-based investigations can be found in a number of different places. These include:

- 'Project pages' from past copies of *Chemistry Review* published by Philip Allan

- www.york.ac.uk/chemistry/ schools/chemrev/projects/

- Salters Advanced Chemistry website: www.york.ac.uk/org/ seg/salters/chemistry

- The *School Science Review* (Association for Science Education): www.ase.org.uk/ journals/school-science-review

- The RSC (Royal Society of Chemistry): www.rsc.org/ learn-chemistry

- ILPAC (Independent Learning Project for A-level Chemistry): https://www.stem.org.uk/ elibrary/collection/3844

- University of Canterbury, New Zealand: www.outreach. canterbury.ac.nz/chemistry/ index.shtml

- *Journal of Chemical Education* (JCE): http://pubs.acs.org/journal/ jceda8/

You may design and build your own apparatus, such as a polarimeter for investigating optical activity in solution or for investigating paramagnetism (Figure 6.1).

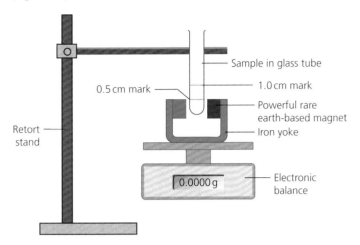

Figure 6.1 One possible apparatus for measuring paramagnetism in a transition metal salt sample

Personal engagement criterion checklist

Creativity input and initiative

Descriptor	How?
You demonstrate creativity during the investigation.	
You use a suitably modified and referenced methodology.	
You demonstrate independent thinking during the entire investigation.	
You demonstrate initiative during the entire investigation.	

Justification for research question

Descriptor	How?
You justify why you chose to investigate your research question and include your personal interest and curiosity.	
You discuss the wider importance/impact of the research question.	

Exploration

This criterion assesses the extent to which you establish the scientific context for your work, state a clear and focused research question and use concepts and techniques appropriate to diploma level. Where appropriate, this criterion also assesses your awareness of safety, environmental and ethical considerations.

Mark	Descriptor
0	The student's report does not reach a standard described by the descriptors below.
1–2	The topic of the investigation is identified and a research question of some relevance is stated but it is not focused.
	The background information provided for the investigation is superficial or of limited relevance and does not aid the understanding of the context of the investigation.
	The methodology of the investigation is only appropriate to address the research question to a very limited extent since it takes into consideration few of the significant factors that may influence the relevance, reliability and sufficiency of the collected data.
	The report shows evidence of limited awareness of the significant safety, ethical or environmental issues that are relevant to the methodology of the investigation.*
3–4	The topic of the investigation is identified and a relevant but not fully focused research question is described.
	The background information provided for the investigation is mainly appropriate and relevant and aids the understanding of the context of the investigation.
	The methodology of the investigation is mainly appropriate to address the research question but has limitations since it takes into consideration only some of the significant factors that may influence the relevance, reliability and sufficiency of the collected data.
	The report shows evidence of some awareness of the significant safety, ethical or environmental issues that are relevant to the methodology of the investigation.*
5–6	The topic of the investigation is identified and a relevant and fully focused research question is clearly described.
	The background information provided for the investigation is entirely appropriate and relevant and enhances the understanding of the context of the investigation.
	The methodology of the investigation is highly appropriate to address the research question because it takes into consideration all, or nearly all, of the significant factors that may influence the relevance, reliability and sufficiency of the collected data.
	The report shows evidence of full awareness of the significant safety, ethical or environmental issues that are relevant to the methodology of the investigation.*

*Methodology includes all the procedures, protocols, and techniques for collecting and analysing experimental data.

© IBO 2014

Table 7.1 Mark descriptors for the exploration criterion

A suitable individual investigation may involve studying how an independent variable (numerical and continuous) affects a single dependent variable, while a range of other variables are controlled or at least monitored. In addition to a focused research question, this criterion will assess background information that provides context and reasons for the investigation. This needs to be focused and contain only relevant information. The independent variable and its range need to be stated and justified, and all other variables identified with their methods of measurement. Trials may be used to determine appropriate values for an independent variable. Your discussion of controlled variables should show that you understand how other factors may impact on values of the dependent variable (fair test). The internal assessment report needs to describe and explain the safety, ethics and environmental impact of the investigation.

Examiner guidance

A synthesis, extraction and separation investigation may generate raw data that cannot be meaningfully analyzed or interpreted. Having discrete data (such as brands) rather than continuous data could make it difficult to gain maximum marks for processed data.

Worked example

A 'potato' cell (Figure 7.1) may be made by inserting copper and zinc electrodes into a potato. The electrolyte is a weak acid dissociated into hydrogen ions (H^+) and conjugate base (A^-). Connections can then be made to an external circuit and the cell potential measured. The internal resistance of the potato cell is measured by a resistance box (ohmmeter) and values of resistance are calculated from the current produced and the voltage developed across the resistor.

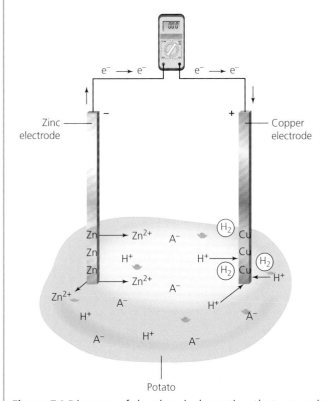

Figure 7.1 Diagram of the chemical reaction that occurs in a zinc/copper potato cell, with a multimeter connected between the electrodes

The investigation might also determine the maximum power output (voltage × current) obtainable from the potato cell but these are potential dependent variables and just give single values.

Ideas for investigations involving a 'potato' cell

- Investigate the effect of electrode distance.
- Investigate the effect of temperature (that lowers the internal resistance).
- Investigate how long the potato (or equivalent) can sustain a current and power an electronic clock.

The investigation should only have one independent variable and issues with biological variation should be recognized. The potato can be replaced with fruits with higher electrolyte concentration. Comparing the voltage of a potato, an orange and a lemon is not a good choice of an independent variable. The change in the pH of a given fruit (via degree of ripening) would be a better independent variable to study.

The potato cell will have a limited power output due to the production of hydrogen gas (that will eventually reach a steady state) at the copper electrode; at the same time, the zinc electrode will acquire deposits of oxides that act as a barrier between the metal and the electrolyte (polarization). A long-term limitation will be loss of zinc or depletion of chemicals present in the potato.

Worked example

It is advisable to find out how a measureable factor will affect the yield or purity of a chemical synthesis, such as aspirin. The effect on the yield or purity of the different reagents (for example, phosphoric(v) versus sulfuric acid) could be investigated. The effect of reaction time, temperature, reagent ratios, and so on, would also give a numerical independent variable. Data collected from experiments, by you, is **primary data**.

Research question

The research question **may** be phrased as a question. Consider an investigation involving urease (that can be purchased or extracted from soya beans). The hydrolysis of urea results in the formation of ammonia and carbon dioxide, causing the pH to rise as the reaction proceeds.

Research question:

> How is the reaction rate of hydrolysis of urease affected by the presence of copper(II) ions in aqueous solution as measured by the change in pH?

■ ACTIVITY

1   List other possible research questions for the enzyme-controlled hydrolysis of urea in aqueous solution.

A focused research question typically identifies the independent variable, the dependent variable and perhaps the method for a well-defined chemical system. It does **not** have to be phrased as a question. For example:

> To investigate the relationship between the rate at which banana extract turns brown (due to polyphenol oxidase action) as measured by absorbance at 450nm and pH7.

The independent variable is pH (hydrogen ion concentration) and the dependent variable is the rate of reaction, which may be measured in various units. It might be helpful to specify the unit of rate and its type (presumably average) and it may be acceptable just to refer to colorimetry. The details could be present in the background information and methodology.

■ ACTIVITY

2   Outline why the following research question is unsuitable and what is missing:

> To investigate the kinetics of polymerization of ethanal in the presence of alkali.

> **Examiner guidance**
>
> **Key variables** are independent variables with a large effect on a dependent variable, for example, temperature on rate of chemical reactions.

Background information

A review of the chemical literature (which must be fully referenced) related to your individual investigation has the following functions:

■   to justify your choice of research question and methodology

■   to establish the importance and significance of the chemical topic

■   to provide the relevant background information needed to understand your report.

Your background information should provide a brief overview of the theory and current knowledge, with an emphasis on the chemical literature specific to your topic. It should support the argument behind your report, using evidence from that research area, but should not include any advanced chemical theory that is not directly relevant or that you do not understand and cannot explain clearly.

Constructing a hypothesis

You **could** use your research question to formulate a hypothesis, and this **might** be a useful inclusion in your background information, but it is not an explicit requirement of the group 4 assessment criteria.

A hypothesis is a testable prediction and explanation of the type of chemical behaviour (in terms of chemical models and theories) or results expected during a chemical investigation. It usually involves a causal relationship between an independent and dependent variable and it should be either tentatively supported or falsified by your experiment.

For example, you could analyze the results of kinetics experiments in terms of simple collision theory and transition state theory (models), the Maxwell–Boltzmann distribution and concepts such as activation energy and rate determining step. You can assume ideal behaviour of the particles in the gas phase or dilute solution.

Variables

Consider the research question:

> What is the effect of fluoride ion concentration on the kinetics of the reaction between calcium carbonate and dilute hydrochloric acid?

The independent variable is the concentration of fluoride ions and the dependent variable is the rate of reaction (possibly measured by total carbon dioxide volume after a fixed time interval). The relevant controlled variables are temperature and concentration of the acid, surface area and mass of calcium carbonate.

The variables could be classified and tabulated with an emphasis on explicitly identifying the controlled variables that affect the results. Table 7.2 shows one suggested format for the presentation of information about variables.

Type of variable	Variable	Method for control	Reason for control

Table 7.2 A possible format outlining the classification of variables

Methodology

Your investigation must record **relevant**, **reliable** and **sufficient** raw data to address the research question. Relevant means the data is related to your research question; reliable means precise and accurate; and sufficient means a wide range of repeated data is collected. You should plan and perform enough replicates to establish the basic **reproducibility** of your method, and a wide range / sufficient values of your independent variable.

For example, an investigation to find out if the rate of a chemical reaction depends upon the concentration of one of the reactants would not be a relevant procedure if you did not control the temperature of the reactants.

You should give a detailed method with a labelled diagram drawn in cross section or a labelled digital photograph. You should limit your drawings to complex set-ups, non-standard equipment or standard equipment being used in an unusual manner. There is no need to document (by including diagrams or photographs of burettes or pipettes) a standard titration or dilution in great detail.

You should provide enough detail so that another student could repeat your work. This means you should state actual quantities of substances and how you will measure out these quantities; concentrations; and sizes and precision of apparatus and instruments.

You can write your method in continuous prose or as a list (in bullet points). You should describe clearly any steps or procedures you design to minimize the systematic and random errors in all of your measurements. Your method should also deal with any limiting reagent and excess reagent issues.

If you need to control the temperature (via a water bath with a thermostat), this should be measured frequently. In investigations in which temperature needs to be controlled, such as reaction rates or equilibrium, it is the temperature of the reacting chemicals that is important, and not the temperature of the laboratory.

Where appropriate, you should outline briefly the advantages and limitations of one type of apparatus or instrument, control measure or practical approach compared to other possibilities. It may be helpful to have two sections: your plan or design and development of the method, and the methodology you actually used.

Safety, ethics and environmental impact

Your internal assessment report should consider safety, ethical and environmental issues. Outline how reagents should be stored, handled safely and disposed of (with minimum impact on the environment). If there are no serious issues to be addressed, you should state there are **no** safety/environmental issues. You should also consider sustainable consumption of chemicals (for example, use of microscale synthesis).

Risk assessment

Your internal assessment report must contain a risk assessment where relevant (it is not required for a simulation). The three main parts of a risk assessment are:

- **Hazard identification**: identifying safety and health hazards associated with laboratory work

- **Risk evaluation**: assessing the risks involved

- **Risk control**: using risk control measures to eliminate the hazards or reduce the risks.

Hazard labelling systems

A 'hazard' is any source of potential harm to an individual's health under certain conditions in the laboratory. 'Risk' is the probability of a person being harmed or experiencing an adverse health effect if exposed to a hazard. For example, exposure of the skin to sodium hydroxide is a hazard, and blistering of the skin is a potential source of harm.

Risk assessment is the process of estimating the probability of harm from a hazard (the severity of the hazard multiplied by the probability of exposure to the hazard), by considering the process or the laboratory procedure that will be used with the hazard.

Carrying out a risk assessment involves estimating the risk and then identifying steps to minimize the risk, for example, reducing the quantity of the hazard being handled, using chemical fume hoods, plastic safety screen, devising safe procedures for handling the hazard, and wearing safety glasses and laboratory coat.

Classifying hazardous chemicals

The Globally Harmonized System (GHS) is an internationally adopted system from the United States for the classification and labelling of hazardous chemicals. The GHS provides established description and symbols (Figure 7.2) for each hazard class and each category within a class. This description includes a signal word (such as 'danger' or 'warning'), a symbol or pictogram (such as a flame within a red-bordered diamond), a hazard statement (such as 'causes serious eye damage'), and precautionary statements.

Figure 7.2 Hazard warning signs

Safety Data Sheets

The Safety Data Sheet (SDS) is provided in the US by the manufacturer, distributor, or importer of a chemical, to provide information about the substance and its use. The information includes the properties of each chemical; the physical, health, and environmental health hazards; protective measures; and safety precautions for handling, storing, disposing of, and transporting the chemical.

Before an experiment

Carefully develop a list of all of the chemicals used and the quantities needed in an experiment. You should always identify the substance you are working with and think about how you can minimize exposure to it during the experiment.

Find and evaluate hazard information from the SDS, which suppliers are required to provide the end user. The label of the original container also contains valuable safety information.

Determine the minimum quantity of each chemical or solution that will be required for completion of an experiment. Build in a small excess, avoiding having large excesses that will require disposal.

Ensure that the proper concentrations are prepared and all chemical bottles are properly labelled with name, formula, concentration and any hazard warnings (such as corrosive or caustic, and its pictogram).

Disposal of waste

Table 7.3 shows how chemical waste may be properly disposed of in a laboratory. You must consult your teacher and laboratory technician.

> **Expert tip**
>
> Always read the label on a chemical reagent bottle to obtain and review basic safety information concerning the properties of a chemical. It is your responsibility, in conjunction with your teacher, to be fully aware of the hazards and risks of all chemicals being used.

> **Expert tip**
>
> 'Exposure limit' is the established concentration of a chemical that most people could be exposed to in a typical day, without experiencing adverse effects. Exposure limits help you to understand the relative risks of chemicals.

> **Examiner guidance**
>
> Common laboratory hot plates are not designed for the heating of flammable or combustible chemicals. A burner should not be used to heat a flammable or combustible chemical. If flammable substances need to be heated, this should be done in small quantities, for example, in a hot water bath and in a fume hood.

Aqueous waste	Organic chemicals	Solid waste	Special case
Acidic (pH < 4)	Non-chlorinated (for example, hexane, menthanol)	Lightly contaminated • No visible loose powders	Sharps (for example, needles, razor blades)
Neutral (pH 4–10)	Chlorinated (for example, chlorobenzene)	Chemical • Loose powders • Heavily contaminated solid materials (for example, used filter paper, unwanted samples, heavily contaminated gloves)	Inorganic oxidizing agents • Place in a container with a disposal label
Basic (pH > 10)	Chemicals in a commercial bottle Undamaged bottle: dispose in original bottle (no label necessary)	Silica gel • Dispose in separate container • May not be combined with other types of chemical waste	Reactive • Consult your lab technician
	Damaged bottle (inform laboratory technician and teacher)	Chemicals in a commercial bottle Undamaged bottle: dispose in original bottle (no label necessary).	
		Damaged bottle (inform laboratory technician and teacher)	

Table 7.3 Correct disposal of chemical waste

When assessing safety, ethics, and environmental issues, you should ensure that the following are considered and included in your IA report:

- evidence of a risk assessment
- an appreciation of the safe handling of chemicals or equipment (for example, the use of protective clothing and eye protection)
- a reasonable consumption of materials
- the use of consent forms in human physiology experimentation
- the correct disposal of waste
- attempts to minimize the impact of the investigation on field sites.

Exploration criterion checklist

Defining the problem and selecting variables

Descriptor	Complete
Research question	
You identify an appropriate topic.	
You state a relevant, specific and fully focused research question (that must not overlap with an extended essay or an internal assessment in another subject).	
Your research question sets your framework (and aim) and this is consistently carried through.	
You clearly identify and consider all the relevant variables.	
You predict, when appropriate, a quantitative relationship between the independent variable and the dependent variable(s).	
You state and describe the relevant controlled variables together with why and how they are controlled or monitored.	
Background information	
You give detailed relevant chemical background theory and information that enhances understanding of your investigation and put it into a chemical context.	
You include a hypothesis or chemical model where appropriate.	
You outline any assumptions or simplifications in any chemical models or theories.	
You include a brief survey or summary of the chemical literature, referenced using a recognized referencing style.	
Safety, ethical and environmental issues	
Your plan shows you are aware of safety, ethical (green chemistry) and environmental issues related to the methodology (for example, risk assessment, use of chemical reagents and solvents, storage and disposal of chemicals).	

Controlling variables

Descriptor	Complete
You include a description of chemical apparatus and instrumentation (including range, sensitivity and absolute uncertainty in a single measurement).	
You include a description of quantity (mass/volume (with units), concentration (with units) of solutions, physical state of solids, and so on). A single reference in the method is acceptable.	
You give a clear, detailed and logical sequence of reproducible steps.	
You describe the rationale or justification of relevant steps in the method.	
You describe how your methodology minimizes random and systematic errors.	
You include a cross-sectional and labelled diagram or annotated photo showing arrangement of non-standard apparatus. You use correct names and terminology.	
You describe how and why controlled variables are to be held constant or monitored (if they cannot be controlled).	
You describe how the independent variable is varied and state the values/range you have chosen for manipulation.	
You describe how the dependent variable is measured and how you will deduce the processed variable(s) from the raw data.	
You outline planned controls (if appropriate) and simple statistics, if you record large numbers of repeated readings.	
You explain choices with regard to the methodology, apparatus, instruments, materials or reagents.	
You explore alternative methods and outline why they are less suitable.	

Planning and recording of data

Descriptor	Complete
You plan to collect a sufficient number of reliable and relevant raw data points over a wide data range.	
You plan to collect more data at certain points, if appropriate (for example, at the extremes of the ranges and inflexion points).	
You plan to collect a suitable number of repeated and averaged readings.	
You plan to collect relevant qualitative data (observations).	
You ensure that your data collection is relevant to the initial research question.	
You plan to collect raw data with units and absolute random uncertainties, recorded to an appropriate precision.	
You plan to record physical conditions such as temperature and pressure, if these affect the value of your dependent variable(s).	

8 Analysis

This criterion assesses the extent to which your report provides evidence that you have selected, recorded, processed and interpreted the data in ways that are relevant to your research question and can support a conclusion.

Mark	Descriptor
0	The student's report does not reach a standard described by the descriptors below.
1–2	The report includes insufficient relevant raw data to support a **valid conclusion** to the research question. Some basic data processing is carried out but is either too inaccurate or too insufficient to lead to a valid conclusion. The report shows evidence of little consideration of the impact of measurement uncertainty on the analysis. The processed data is incorrectly or insufficiently interpreted so that the conclusion is invalid or very incomplete.
3–4	The report includes relevant but incomplete quantitative and qualitative raw data that could support a simple or partially valid conclusion to the research question. Appropriate and sufficient data processing is carried out that could lead to a broadly valid conclusion but there are significant inaccuracies and inconsistencies in the processing. The report shows evidence of some consideration of the impact of measurement uncertainty on the analysis. The processed data is interpreted so that a broadly valid but incomplete or limited conclusion to the research question can be deduced.
5–6	The report includes sufficient relevant quantitative and qualitative raw data that could support a detailed and valid conclusion to the research question. Appropriate and sufficient data processing is carried out with the accuracy required to enable a conclusion to the research question to be drawn that is fully consistent with the experimental data. The report shows evidence of full and appropriate consideration of the impact of measurement uncertainty on the analysis. The processed data is correctly interpreted so that a completely valid and detailed conclusion to the research question can be deduced.

© IBO 2014

Table 8.1 Mark descriptors for the Analysis criterion

Recording and presenting raw data

Tabulating data

Your internal assessment will involve accurately recording qualitative data (observations, for example, chromatograms), and quantitative data (measurements), often in data tables (Table 8.2) with headings containing units and absolute random uncertainties. The decimal places used for each variable must be consistent with the precision of the apparatus or instrument you use. You must use the same number of decimal places consistently across all data for each variable.

Temperature/°C ± 0.05 °C	Pressure/Pa ± 0.1 Pa
73.50	97 991.9
65.40	72 260.7
59.75	61 328.3
59.95	49 062.6
40.45	37 330.3
30.75	28 664.3

Table 8.2 The vapour pressure of carbon tetrachloride, $CICl_4(g)$, at different temperatures

The independent variable should be in the left-hand column in a table, with the following columns showing the dependent variable and then any processed variables. The body of the table should not contain units. Data in the columns are generally arranged in order of increasing or decreasing values of the independent variable (for example, the recording of raw titration data is shown in Chapter 4).

You can list data consisting of very large or very small numbers in scientific notation in one of the three ways shown in Table 8.3. In the second column, the actual values have been multiplied by the value given in the heading to produce data in a convenient form, while the third column indicates that values of $10^{-3}\,mol\,dm^{-3}$ are being recorded. This approach may also be appropriate for graphs (for example, an Arrhenius plot).

Concentration/mol dm^{-3} (±0.1 × 10^{-3})	Concentration/mol dm^{-3} × 10^{-3} (±0.1)	Concentration/10^{-3} mol dm^{-3} (±0.1 × 10^{-3})
1.1×10^{-3}	1.1	1.1
1.2×10^{-3}	1.2	1.2
1.3×10^{-3}	1.3	1.3

Table 8.3 Raw data related to concentrations

You should cite properly any secondary data you take from a literature source. You should indicate any unreliable data using an asterisk.

Examiner guidance

If you have a relatively large amount of numerical raw data, it may be acceptable to present one set of raw data for one trial and averaged data for the other trials. When large amounts of primary raw data have been collected, using data-logging, for example, you should only present a representative sample of the raw data.

If an experiment results in poor data, you should consider the need to modify the experimental conditions to improve the data (for example, low titres from a titration or where a gas is evolved too quickly or too slowly during a kinetics investigation). You should record this preliminary unsatisfactory data in your internal assessment report and comment on it in your evaluation. The modifications you make will be considered under 'personal engagement'.

Recording observations

Where appropriate, you must record observations or qualitative data (for example, indicator colour changes, precipitates forming, gases released, colours of solutions or precipitates, shapes and colours of crystals formed and changes in smell, noises). However, you should only record observations where they are relevant to allow the research question to be answered.

Examiner guidance

Do not use words such as 'see-through', 'clear' or 'transparent', when you mean 'colourless'. You can 'see through' dilute potassium dichromate(VI) solution, but it is not colourless, and copper(II) nitrate solution is 'transparent' but not colourless.

Data processing

You need to carry out complete and correct quantitative processing of the raw data. Ideally, you should carry out a range of mathematical processing on averaged data. You should give one sample calculation with all the steps shown; other identical calculations of the same type need not be shown. An example of poor data processing would be the inclusion of a bar chart with no associated calculation.

In numerical calculations, intermediate results carry one or two extra digits beyond the last significant one, but you must quote the raw data and your final results with the correct number of significant figures (see Chapter 4) and units.

The number of significant figures in the final calculated result depends on the apparatus you used and the accuracy of your measurements. This is usually the same as the lowest number of significant figures in any measurement you used to determine your final result (Figure 8.1).

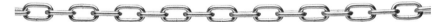

Figure 8.1 A chain is as strong as its weakest link. A calculated answer is as accurate as the least accurate measurement in the calculation.

Examiner guidance

If you have large amounts of raw data from the analysis of a biological material then it may be appropriate to use some simple statistics (for example, range and standard deviation) or a statistical test (for example, a *t*-test, assuming a Gaussian distribution).

Presenting data: graphing

Adding a straight line to a scatter graph implies a linear correlation between the two variables. The points on a line graph are connected by a best fit line, which is a line drawn so that as many points fall above the line as below it, and such that the points are evenly distributed on either side of the line.

Examiner guidance

It is helpful if you include error bars on line graphs with maximum and minimum slopes for linear graphs. You should also consider the equation of any straight line, plus its correlation coefficient (R^2 value) (see Chapter 5).

Impact of measurement uncertainty

Propagate uncertainties (Chapter 4) through a calculation by using the absolute and/or percentage uncertainties from measurements to determine the overall uncertainty in calculated results. For functions such as addition and subtraction, absolute uncertainties can be added. For multiplication, division and powers, percentage uncertainties can be added. If one uncertainty is much larger than others, the overall uncertainty in the calculated result can be taken as due to that quantity alone.

Expert tip

It is suggested you draw a graph of the results while the experiment is in progress. Any data point not conforming to the trend can then be identified and re-assessed by immediately recording the measurement again. You should **not** discard data points that do not conform but discuss them in your evaluation. This may also help you to modify your method, which is recognized under 'personal engagement'.

Common mistake

Graphs are sometimes reduced in size to ensure the report is the correct page length. Graphs should not be reduced to such a size that they become uninformative, simply to stay within the page limit. It is also important to choose a suitable graph style so information can be drawn from it.

You can compare this overall percentage error with different data sources (secondary data examples) to assess the variance. This is the square of the standard deviation (see Chapter 4) and a measurement of how far random numbers are spread out from their mean.

You can use gradients of minimum and maximum lines for a line graph with error bars to determine the random uncertainty. You should omit outliers with appropriate consideration (ideally, statistically, using a Q-test or two standard deviation test.

RESOURCES

Visit the website 'Dixon's Q-test: Detection of a single outlier' for more information on detecting a single outlier:

https://tinyurl.com/y9ltvtsg

Interpreting processed data

This is the process of making chemical and mathematical sense of the data that will lead to your conclusion. You should explain trends in graphs with reference to: optima (peaks), maxima or minima (plateaus), gradients (zero, positive or negative and increasing/decreasing) or intercepts (on the x- or y-axis), and identify their mathematical relationships (such as linear, directly proportional, inversely proportional, negative or positive exponential) from their characteristic shapes (Chapter 4). If you have performed statistical calculations then you should outline their significance. Where possible and appropriate, you should linearize nonlinear relationships.

Analysis criterion checklist

Recording raw data

Descriptor	Complete
You neatly record all raw data: qualitative data (observations) and quantitative (numerical) data necessary to support a conclusion to your research question.	
You present all raw numerical data clearly and correctly in tabulated manner (for example, independent variable on the far left, then dependent variable followed by processed variables). SI units are typically used.	
Your data tables have labels, units and uncertainties (absolute errors) once, in the headings.	
You consistently record quantitative data, taking into account the absolute uncertainty/error (correct number of decimal places).	
If using scientific notation, you quote the value and the error with the same exponent.	
You describe how the absolute uncertainties in the measurements were obtained (for example, half of the least count of the scale, smallest digital reading or manufacturer's tolerance).	
You mention the reaction time for manual timings.	
You clearly indicate and highlight any anomalous data (and statistically justify its exclusion, if appropriate).	
Your processed data resolves the research question.	

Processing raw data

Descriptor	Complete
You use the appropriate mathematical equations to carry out calculations to process raw data.	
You record results of calculations with the correct number of significant figures.	
You convert the averaged data into the correct graphical form: line graph, scatter graph, bar chart and so on.	
You select processed average data to produce a straight line graph (if data is continuous) (with line of best fit).	
Your graphs include the equation of the line and the R^2 value, where appropriate.	
You extract relevant quantities from the line graph (for example, gradients, intercept) or perform interpolation or extrapolation.	
You process replicated data, finding the mean and using the variation between values to assign an appropriate uncertainty.	

Presenting raw data

Descriptor	Complete
You use scientific conventions in tables of processed data.	
You use accepted conventions for graphs (for example, title, correct graph size (large), appropriate range and scale, labelling, units, uncertainties).	
Your graphs have the independent variable on the x-axis and the dependent variable on the y-axis.	
You draw the line or curve of best fit correctly and clearly indicate the trend shown.	
You include error bars on line graphs, where appropriate and possible.	

Impact of uncertainty

Descriptor	Complete
You convert absolute uncertainties to percentage errors and propagate uncertainties appropriately and consistently.	
You present calculations and error propagation in a clear, organized and separate manner.	
You present final figures with the number of decimal places consistent with the propagated uncertainty.	

Interpreting processed data

Descriptor	Complete
The processed data is correctly interpreted (for example, the correct graphical relationship between the independent and the processed dependent variable is deduced from a graph with data quoted to support it).	

9 Evaluation

This criterion assesses the extent to which your report provides evidence of evaluation of the investigation and the results with regard to the research question and the accepted scientific context.

Mark	Descriptor
0	The student's report does not reach a standard described by the descriptors below.
1–2	A conclusion is outlined which is not relevant to the research question or is not supported by the data presented.
	The conclusion makes superficial comparison to the accepted scientific context.
	Strengths and weaknesses of the investigation, such as limitations of the data and sources of error, are outlined but are restricted to an account of the practical or procedural issues faced.
	The student has outlined very few realistic and relevant suggestions for the improvement and extension of the investigation.
3–4	A conclusion is described which is relevant to the research question and supported by the data presented.
	A conclusion is described which makes some relevant comparison to the accepted scientific context.
	Strengths and weaknesses of the investigation, such as limitations of the data and sources of error, are described and provide evidence of some awareness of the methodological issues involved in establishing the conclusion.
	The student has described some realistic and relevant suggestions for the improvement and extension of the investigation.
5–6	A conclusion is described and justified which is relevant to the research question and supported by the data presented.
	A conclusion is correctly described and justified through relevant comparison to the accepted scientific context.
	Strengths and weaknesses of the investigation, such as limitations of the data and sources of error, are discussed and provide evidence of a clear understanding of the methodological issues involved in establishing the conclusion.
	The student has discussed realistic and relevant suggestions for the improvement and extension of the investigation.

Table 9.1 Mark descriptors for the Evaluation criterion.

Conclusion

You should interpret observations and measurements and draw conclusions from processed data to answer the research question. Your conclusion should make use of the chemical concepts and ideas you described in your background knowledge, and contrast this information with your outcome. Your conclusion should focus on how the independent variable causally affects the dependent variable. If you have proposed a hypothesis then you should conclude whether the data supports it or does not support it.

You must establish whether your trends or relationships (shown by your graphs) are consistent with the accepted theory and literature (referenced). If you are unable to do this, then you need to demonstrate that your raw data is reliable and that your data processing is appropriate.

You should compare the experimental accuracy (the percentage difference between experimental and literature values) with the total random error (the final propagated uncertainty) and comment on it.

If the experimental error is greater than the random error then random error alone does not explain the difference between your value and the literature value. This indicates systematic errors.

Common mistake

Evaluation is often the weakest criterion in a report. It is a difficult skill to learn and apply, but sometimes candidates finish off the report in a hurry, leading to an inadequate evaluation, or devote insufficient pages to this section, compromising its quality.

Examiner guidance

Even if your experimental error is smaller than your random error, you should still discuss systematic errors since they might have cancelled each other out.

Worked example

A eudiometer (Chapter 1) was filled with hydrogen from the reaction between magnesium and hydrochloric acid. The pressure, volume and temperature were measured directly, but the amount of hydrogen was calculated from the amount of magnesium via the equation. These results were used to calculate an experimental value for the gas constant, R.

$P = 9.77 \times 10^4 \, \text{Pa} \pm 80 \, \text{Pa}$
$V = 3.63 \times 10^{-4} \, \text{m}^3 \pm 2.0 \times 10^{-6} \, \text{m}^3$
$n = 0.01470 \, \text{mol} \pm 0.00015 \, \text{mol}$
$T = 298.8 \, \text{K} \pm 0.2 \, \text{K}$

Converting these absolute random uncertainties to percentage uncertainties:

$P = 9.77 \times 10^4 \, \text{Pa} \pm 0.08 \, \%$
$V = 3.63 \times 10^{-4} \, \text{m}^3 \pm 0.55 \, \%$
$n = 0.01470 \, \text{mol} \pm 1.02 \, \%$
$T = 298.8 \, \text{K} \pm 0.07 \, \%$

The total random uncertainty = 1.72 %.

$$R = \frac{pV}{nT}$$

$$= \frac{(9.77 \times 10^4 \, \text{Pa} \times 3.63 \times 10^{-4} \, \text{m}^3)}{(0.0147 \, \text{mol} \times 298.8 \, \text{K})}$$

$$= 8.07 \, \text{J} \, \text{mol}^{-1} \, \text{K}^{-1}$$

The percentage difference between the experimentally determined value and the data book value: $[(8.31 - 8.07) \div 8.31] \times 100 = 2.89 \, \%$, so the experimental inaccuracy = 2.89 % = 2.9 %. The value of R is −2.89 % as the experimental value is below the literature value.

Since the experimental inaccuracy is greater than the total random percentage uncertainty, there must be a systematic error(s) in the investigation, such as leakage of hydrogen gas from apparatus or a slight dissolution of the gas in the water, and, more significantly, oxide formation on the surface of the magnesium. The partial pressure of water vapour was also not taken into account.

Examiner guidance

The direction of a systematic error on the measured value should be included.

Strengths and weaknesses of the investigation

The methodology must be evaluated thoroughly. You need to include the strengths as well as the weaknesses of the investigation. For example, a strength could be a small range or standard deviation in data for the dependent variable, indicating that repeated measurements were precise. A weakness may be lack of control of some uncontrolled variables. Do not restrict weaknesses to details of the practical, but include other aspects as well (for example, lack of secondary data or literature values for comparison).

Limitations of the data and sources of error

In this part of the report you can discuss uncertainties (percentage error) of dependent and independent variables. The impact of the limitations on the conclusion must be discussed. Your report needs to propose improvements, and these must be realistic and specific.

You should refer to systematic and random errors with the relative magnitude of each. The total percentage random error should be calculated and the size and effect of probable systematic errors noted (where possible). Factors to consider include **validity of the method**, such as the range (of the dependent variable), sample size, sampling method (if appropriate) and the use of an alternative reaction system to study the same chemical phenomenon.

Examiner guidance

You should only consider significant sources of error. For example, in a calorimetry experiment using spirit burners to burn alcohols in air, the major sources of (systematic) error are incomplete combustion and heat loss to the surroundings, and these are far more significant than any random errors in the experiment. Discussing in detail the random errors in masses, volumes or temperatures is not necessary and will take up valuable space in your report.

Worked example

Consider an investigation into the effect of aqueous solutions of halogens on the relative rates of reaction of the halogenation of propanone as measured by the time (measured manually) for the colour to disappear upon manual shaking of the reactants.

Bromine and iodine react with the halide ion to form a colourless trihalide ion, for example:

$$I_2(aq) + I^-(aq) \rightarrow I_3^-(aq)$$

This will distort the kinetics and colours observed. This is a systematic error of known magnitude (if the equilibrium constants are located or experimentally measured) that will reduce the rate and make times longer. The shaking was not accurately controlled and hence introduces a small systematic error.

Improvements and extensions

Your suggested improvements should be precise, focused and relevant to your investigation. You should relate them directly to the weaknesses or limitations you identify and it should be feasible to carry these out in a school or university laboratory. A tabulated approach may be helpful (for example, Table 9.2).

Methodology limitation	Impact on results	Suggested improvement
Some of the carbon dioxide collected by displacement of water in the eudiometer may have reacted and dissolved in the water to form carbonic acid.	The volume of carbon dioxide measured is less than the volume released by the reaction.	Collect the gas over hot water or over a hydrocarbon solvent to reduce its solubility.

Table 9.2 Suggested table for evaluation of methodology

The extensions should follow on from your research findings, and, if performed, they should enhance your understanding. Extensions should not just be 'more of the same' (that is, simply repeating the same experiment with a greater frequency of independent or dependent variable measurements); instead, a significant extension is required, such as a different dependent variable, a different independent variable or a different methodology to address the same research question.

■ **ACTIVITY**

Consider the following **outline** of an experiment that used mass measurements and stoichiometric principles to deduce the equation for the reaction between iron and a solution of copper(II) sulfate.

Coarse iron powder (of known mass) was reacted in a ceramic crucible with excess copper(II) sulfate solution (of known concentration at 25 °C) to form copper, which was removed by filtration and dried in an oven at 105 °C for 2 minutes. The mass of the copper was weighed with a balance (±0.01g).

1   Identify weaknesses, major errors and limitations and state how they can be overcome or minimized by changing the methodology.

Evaluation checklist

Concluding

Descriptor	Complete
You include a detailed conclusion (for example, trend between independent and dependent variables) citing numerical values relevant to your research question.	
Your conclusion is supported by your raw and processed data (typically graphs) and the observations.	
You include a conclusion that is described, justified and compared to the relevant chemical literature (if available).	
You identify and comment on any anomalous data.	
Your conclusion is based on your results and uses tentative words such as 'indicate', 'suggest', 'appear to suggest' and 'support'; **not** 'prove'.	
You include a valid conclusion relevant to the research question and that is within the limits of random uncertainties.	
You justify your results with reference to relevant chemical laws and models, theories and principles.	
You discuss limitations to your results.	

Evaluating procedures: strengths and weaknesses

Descriptor	Complete
You describe assumptions that were made which have affected the accuracy of the results (for example, no competing side reactions or ideal behaviour).	
You calculate the total experimental error (propagated random uncertainty) and compare this to the literature value (if available).	
You identify systematic errors and their directional effect.	
You compare the percentage errors of each type of data and identify the major error(s).	
You discuss any limitations of the method (for example, limited data range, lack of data around inflexion points, limited instrument sensitivity and small sample size).	

Improving and extending the investigation

Descriptor	Complete
You suggest appropriate modifications to improve the accuracy, precision and reliability of your results (by reducing random and systematic errors) or better control/monitoring of controlled variables.	
You suggest an alternative method or different instrumentation to obtain more accurate data.	
You discuss how the suggested improvements or modifications would improve the reliability, precision and accuracy of your results.	
You propose realistic and relevant extensions to the study (for example, greater range of data, more data from around an inflexion point, new data processing / data presentation, choice of a new independent variable).	
You focus your suggested improvements on the existing research questions; you focus your extensions on the new research questions.	

10 Communication

This criterion assesses whether the investigation is presented and reported in a way that supports effective communication of the focus, process and outcomes.

Mark	Descriptor
0	The student's report does not reach a standard described by the descriptors below.
1–2	The presentation of the investigation is unclear, making it difficult to understand the focus, process and outcomes.
	The report is not well structured and is unclear: the necessary information on focus, process and outcomes is missing or is presented in an incoherent or disorganized way.
	The understanding of the focus, process and outcomes of the investigation is obscured by the presence of inappropriate or irrelevant information.
	There are many errors in the use of subject specific terminology and conventions.*
3–4	The presentation of the investigation is clear. Any errors do not hamper understanding of the focus, process and outcomes.
	The report is well structured and clear: the necessary information on focus, process and outcomes is present and presented in a coherent way.
	The report is relevant and concise thereby facilitating a ready understanding of the focus, process and outcomes of the investigation.
	The use of subject specific terminology and conventions is appropriate and correct. Any errors do not hamper understanding.

* For example, incorrect/missing labelling of graphs, tables, images; use of units, decimal places. For issues of referencing and citations, refer to the academic honesty section.

© IBO 2014

Table 10.1 Mark descriptors for the Communication criterion

Structure and clarity

The presentation of your internal assessment report must be coherent and relevant to the focus (the research question), the process (methodology) and the outcomes (results and conclusion). It should resemble a scientific paper and there should be headings and sub-headings to give a logical sequence. Diagrams and digital images should be used to enhance understanding. There should be a logical flow allowing the IB examiner to understand your thought processes throughout the report. Your methodology should be detailed enough for the experiments to be reproducible, but simplistic and well-known and assumed aspects of your method need not be made explicit. An appendix should not be included since it will reduce your available word count, but it will not be automatically marked down provided the internal assessment report is relevant and concise.

Relevance and conciseness

Your word-processed report (of 12 pages maximum) must be relevant to the research question. It should be easy to follow the development of your ideas and thoughts from the beginning to the end of your report. It should be concise with no unnecessary or repetitive information. Full calculations for processing of all data and all error propagation are not expected: selected examples will be sufficient, freeing up more space for your conclusion and evaluation.

Terminology and conventions

Pay attention to the use of chemical terminology, scientific conventions (for example, letters in mathematical equations are italicized), units, labelling of tables and graphs, significant figures, random uncertainties and nomenclature (names of chemical compounds).

> **Examiner guidance**
>
> You should not report excessive quantities of raw data from data-loggers.

> **Common mistake**
>
> The following are not necessary:
> - using whole pages for titles or contents
> - presenting blank data tables at the end of the method section.

You should usually use IUPAC names for organic chemicals (for example, 'ethanoic acid' not 'acetic acid') and the Stock system (using oxidation numbers) should always be used for the naming of inorganic compounds (for example, 'iron(II) chloride' not 'ferrous chloride'). Use subscripts, superscripts, parentheses, and symbols with state symbols appropriately in chemical formulas and equations. Do not condense the formulae (for example, use C_2H_5OH, rather than C_2H_6O to represent ethanol since the second formula also represents methoxymethane (CH_3–O–CH_3)).

Your internal assessment report needs to reference material via footnotes, endnotes or in-text citations. You should include a full bibliography at the end of your report. The numbers of decimal places or significant figures should be consistent and correspond to the precision of the raw data (which should include errors).

Examiner guidance

You should normally **not** use the following concentration units: percentage by mass or volume; normality (N) and parts per million (ppm) to report final processed data. All of these concentration units can be expressed as molar concentrations ($mol\,dm^{-3}$), but units of ppm are appropriate for the Winkler method (biochemical oxygen demand).

Define any abbreviations that refer to compounds (for example, HEPES, an organic buffer) and instrumentation (for example, CV, cyclic voltammetry) that are not mentioned in the IB chemistry syllabus. Syllabus abbreviations such as EDTA or IR spectroscopy are acceptable. If you use any organisms in your research (for example, brine shrimp to test for short-term toxicity), then use a binomial name.

You must word process and number the pages in your report. You may embed any equations in the text of your report and format these using the 'Equation editor' tool on your word processor (Figure 10.1)

Common mistake

A common mistake is to confuse weight (newtonmeter) and mass (balance). Weight is a force (due to gravity) and measured in newtons (N). **Relative molecular mass** is a dimensionless number (for example, $M_r(CO_2) = 44.01$), but **molar mass** is a physical quantity with units (for example, $M(CO_2) = 44.01\,g\,mol^{-1}$).

Figure 10.1 Equation editor tool bar in Microsoft Word® running on an i-Mac®

Any letters in the equation should be in italics. For example:

$$P = \frac{nRT}{V} - \frac{0.250\,mol \times 8.31\,J\,K^{-1}\,mol^{-1} \times 300\,K}{6.34 \times 10^{-3}\,m^3} = 9.83 \times 10^4\,Pa$$

All tables, graphs and equations should be introduced by a sentence of explanation and have an explanatory label and sequential reference in the text.

You may draw and embed structural and three-dimensional formulas, mechanisms and apparatus (in cross-section) using ChemDraw® or free software such as ChemSketch®. You could also use structural formulas and other diagrams from the internet but you must suitably reference and acknowledge these. Alternatively, you can draw diagrams by hand and scan them.

Arrows are very important within organic chemistry (Figure 10.2). The wrong arrow can completely change the meaning of an equation.

RESOURCES

ChemSketch® is a useful website for drawing three-dimensional formulas:

http://www.acdlabs.com/resources/freeware/chemsketch/

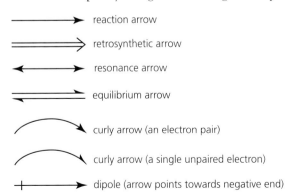

reaction arrow

retrosynthetic arrow

resonance arrow

equilibrium arrow

curly arrow (an electron pair)

curly arrow (a single unpaired electron)

dipole (arrow points towards negative end)

Figure 10.2 Arrows used in organic chemistry

■ SI units

You are normally expected to use the units listed in Table 10.2 for measurements and calculations during practical work for the IB chemistry programme.

Measurement	Unit
Amount	mol (not volume or mass)
Concentration	moles per cubic decimetre ($mol\,dm^{-3}$) or grams per cubic decimetre ($g\,dm^{-3}$)
Energy	joules (J)
Enthalpy change	kilojoules per mole ($kJ\,mol^{-1}$)
Specific heat capacity	joules per gram per degree Celsius ($J\,g^{-1}\,°C^{-1}$ or $J\,g^{-1}\,K^{-1}$)
Entropy change or absolute entropy	joules per kelvin per mole ($J\,K^{-1}\,mol^{-1}$)
Mass	grams (g)
Rate of reaction	moles per cubic decimetre per second ($mol\,dm^{-3}\,s^{-1}$) s^{-1} used for a comparison of rates
Temperature	Thermometers usually measure temperature in degrees Celsius (°C) and often conversion to kelvin (K) is not required; for example, in enthalpy change determinations. However, calculations using the ideal gas equation and the Arrhenius equation require absolute temperatures.
Time	seconds (s)
Charge	coulombs (C)
Potential difference ('voltage') or electromotive force	volts (V)
Current	amperes (or amps) (A)
Volume	Cubic centimetres (cm^3) or cubic decimetres (dm^3). Measurements are usually in cm^3, while concentrations are expressed in terms of dm^3. m^3 may also be used for ideal gas calculations.

Table 10.2 Use of units

Drawing and modelling chemical structures

■ Drawing structures

Drawing the chemical structure of a molecule requires an understanding of the chemical formula, the type of bonding, and often a mental 'picture' of the arrangement of atoms in space.

Figure 10.3 shows the structure of carbon tetrachloride and hexaammine cobalt(II) where standard symbols, wedges and tapers, are used to show the direction of the bonds in three dimensions. This simple notation is used to give a molecule the appearance of a three-dimensional shape.

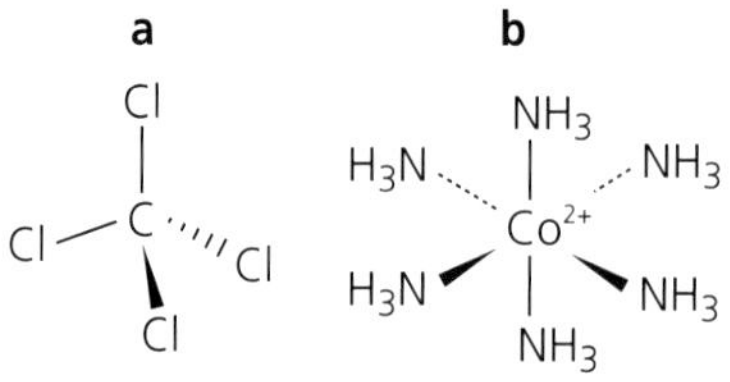

Figure 10.3 Structure of **(a)** carbon tetrachloride and **(b)** hexaammine cobalt(II)

Organic structures are often represented by bond-line formulas which are compatible with computer-based structure drawing packages. The carbon skeleton ('back bone') is represented by a series of lines showing the bonds between the saturated (sp^3 hybridized) carbon atoms but the hydrogen atoms and their bonds to carbon atoms are not shown (Figure 10.4). Stereochemistry can be shown by the usual wedges and tapers (Figure 10.5).

Expert tip

If necessary, show the number of electrons (pairs, or unpaired electrons if radicals) clearly. Remember a '.' (full stop) may be mistaken for a mark in the paper. It may be necessary to show the displayed structural formula of a molecule, for example, in isomerism.

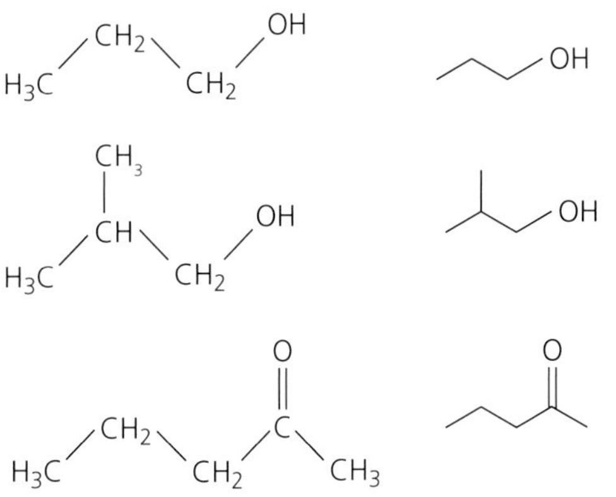

Figure 10.4 Examples of bond-line formulas

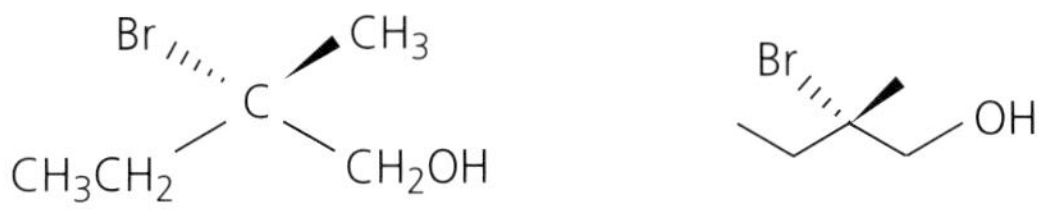

Figure 10.5 Stereochemical representations using bond-line formulas

Internal assessment report format

There is no particular structure that you must follow, but your report should resemble a research paper (without the abstract). If your school does not suggest a format, then you could use the following headings, or simply use the assessment criteria:

- General title or Aim

- Background information (including chemical theory, model and hypothesis (if appropriate))

- Research question

- Risk assessment

- Planning and preliminary work

- Variables

- Methodology

- Raw data

- Processed data (including graphs)

- Error propagation

- Conclusion

- Evaluation

- Random and systematic errors

- Limitations, weaknesses and improvements

- Evaluation of secondary sources

- Future investigations

- Bibliography or References

Examiner guidance

Personal engagement is assessed holistically, so you must show your personal engagement in all other aspects of the report.

■ Using tenses

The introduction is usually in the present tense. For example, *kinetic studies provide important information for reaction mechanisms*. It is usual practice (but not an IB requirement) to use the simple past tense with the passive voice to describe your experiments. For example, *the potassium manganate(VII) solution was standardized using sodium ethanedioate solution*. Results are usually described using the past tense. For example, *three samples of filtered seawater were analyzed with a complexometric titration involving EDTA*.

RESOURCES

You can find out more about using tenses by visiting this website:

http://services.unimelb.edu.au/__ data/assets/pdf_file/0009/471294/ Using_tenses_in_scientific_writing_ Update_051112.pdf

Expert tip

'It's' means 'It is' or 'It has'. The possessive pronoun is 'its', as in 'its experimental value'. It is best to avoid contractions such as it's, hasn't or doesn't in formal scientific writing.

Use Figure 10.6 to check that you have covered all aspects of each criterion.

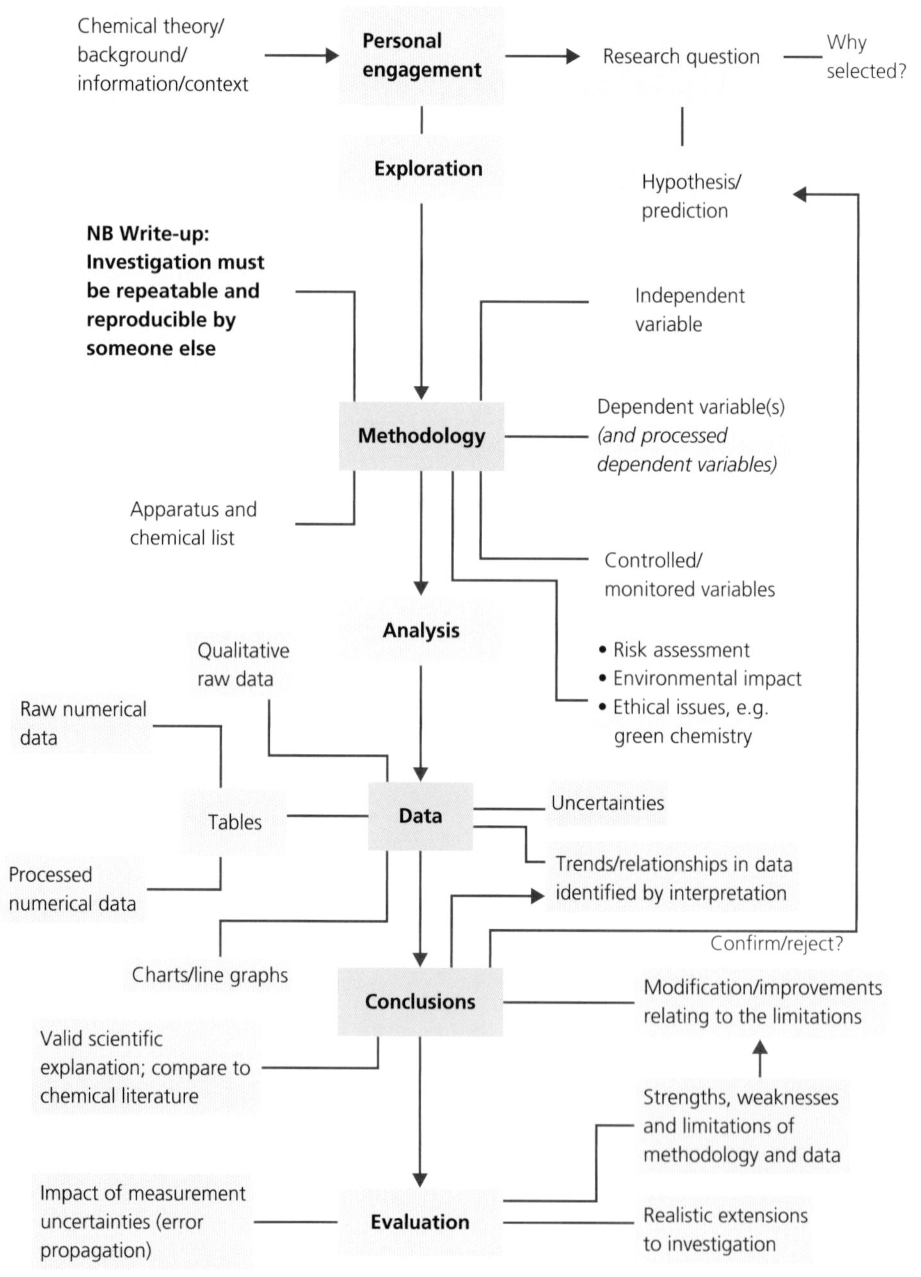

Figure 10.6 Chemistry internal assessment map

Examiner guidance

Communication, like personal engagement, is assessed holistically, meaning the entire report is taken into consideration. A table of contents may give the examiner an overview, but it is not necessary.

Referencing

Referencing is a standardized method of acknowledging the sources of information you consulted. You must reference (acknowledge) words, paragraphs, quotes, figures, images, tables, theories, scientific ideas and facts originating from another source and used in your internal assessment report. Referencing is done:

- to avoid plagiarism
- so that your teacher can verify quotations
- so that your teacher can follow up on your thinking by consulting the source you accessed.

Your school or department is likely to have a referencing style that you must adopt. Be consistent and familiarize yourself with the format and terms that your school or IB chemistry teacher expects you to use.

For example, the ACS style (shown below) has a citation consisting of two parts: the **in-text citation**, that provides brief identifying information within the text, and the **reference list**, a list of sources that provides full bibliographic information.

Talbot, C.D., Harwood, R., Coates, C., *Chemistry for the ID Diploma* (2nd edn), Hodder Education, 2015. (Book)

Cook, A. Gilbert, Tolliver, Randi M., Williams, Janelle E., The Blue Bottle experiment revisited: how blue? How sweet, *J. Am. Chem. Soc*, February 1994, Volume 71, No 2, 160–161. (Journal article)

Woodford, C. Lithium ion batteries. http://science.howstuffworks.com/lead.htm (accessed May 29, 2016). (Website)

Assessing sources

- Can you identify the author's name?
- Can you determine what qualifications or titles he/she has?
- Do you know who employs the author, such as a university or company?
- Is this a primary source (original research paper) or secondary source (review article)?
- Is the content original or derived from other sources?

Evaluating information

It is important that you check the **validity** of the sources you are using: do not assume that information is correct. Apply the following checks:

- Have you checked a range of sources?
- Is the information supported by relevant literature citations?
- Is the age of the source likely to be important regarding the scientific accuracy of the information?
- Is the information scientific fact or opinion and speculation?
- Have the data been analyzed (if appropriate) using relevant statistics?
- Are the data in graphs displayed fairly with error bars?

Academic honesty

You need to ensure that the internal assessment report you submit is your own work. At the end of the course you will need to sign a declaration. Your IB chemistry teacher will check the authenticity of your work, for example, by:

- discussing it with you
- asking you to explain your method and to summarize your results and analysis
- asking you to repeat an investigation
- using software (such as turnitin.com, Figure 10.7) to check for plagiarism.

Expert tip

References are required in both background information and methodology (exploration) and conclusions (evaluation). Have other teachers or researchers carried out similar work to yours that you can refer to? Does your scientific information need referencing?

Expert tip

Scientific papers submitted to peer-reviewed journals such as *Nature* are carefully scrutinized by experts in the field ('peers'). Information from such sources can be trusted as being scientifically valid.

Figure 10.7 Screen shot of turnitin.com

■ Plagiarism

Plagiarism is defined (by the IBO) as the representation, intentionally or unwittingly, of the ideas, words or work of another person without acknowledgement. Academic honesty and integrity is consistent with the IB learner profile (page xvi), where learners strive to be principled. Academic honesty involves values and skills that promote personal integrity and good practice in teaching, learning and assessment.

Examples of plagiarism include:

- copying the work of another IB chemistry student (past or present) or getting someone else to write your report and passing it off as your own work

- copying text or images from a source (book chapter, journal article, or website, for example) and using it without acknowledgement

- quoting the words of others without indicating who wrote or said them (personal communication)

- copying scientific ideas and concepts from a source without acknowledgement, even if you paraphrase them.

Expert tip

The following suggestions will help you to avoid plagiarism:

- Make sure the work (results and theory) you present in your internal assessment report is always your own.
- Never 'copy and paste' from websites or Word® or pdf files downloaded from the internet.
- Place appropriate citations in your report.
- Show clearly where you are quoting directly from a source.

Communication criterion checklist

Descriptor	Complete
Structure of report	
The report is well-structured, coherent and clear, following the style and conventions of a scientific paper.	
Your report is split into appropriate sections (with headings and titles): focus, process and outcomes, the IB criteria or other headings.	
Relevance and conciseness	
Your report only includes relevant information. It is concise and of 6–12 pages in length.	
Your report does not contain any errors, contradictions, false scientific statements or false assertions.	
Subject specific terminology and conventions	
Your graphs, data tables and images are titled and referenced and presented according to conventions in scientific papers.	
Your mathematical equations are in italics and clearly explained and justified/derived (where appropriate) with units.	
You have followed the rules relating to significant figures. The number of decimal places is correct in data tables (that include associated uncertainties) and calculations and error propagation.	
You have defined non-syllabus terms and abbreviations.	
You use correct chemical terms and IUPAC/Stock names.	
Your report includes a cross-referenced bibliography with in-text referencing according to a particular referencing style.	

Glossary

Absolute error – a random error expressed in physical units.

Accuracy – measurements or results that are in close agreement with the true or accepted values.

Accurate measurement – a measurement close to the literature (accepted) value and within its experimental random uncertainty.

Acid–base indicator – an indicator that changes colour on going from acidic to basic solutions.

Analysis/analyze – recognize and comment on trends in raw and processed data and state valid conclusions.

Anomalous data – data with unexpected values that do not match the relationship predicted by the hypothesis. They are not due to random uncertainty in the measurements.

Beer–Lambert law – the linear relationship between absorbance and concentration of an absorbing species.

Calibration – fixing two known points and then marking on a scale on a measuring instrument or establishing two measurements with a digital instrument.

Conclusion – a conclusion is an interpretation based on experimental data.

Control – an experiment where the independent variable is either kept constant or removed. This can be used for comparison, to prove that any changes in the dependent variable in experiments when the independent variable is manipulated must be due to the independent variable rather than other factors.

Controlled variable – a variable that is kept the same in an investigation. In an experiment, at least three controlled variables should be listed, and information about how they will be kept the same included.

Correlation – a statistical measure of the extent to which a linear association exists between two variables.

Data – recorded observations and numerical measurements using apparatus and instruments.

Dependent variable – the dependent variable is the variable whose value is measured (or observed) for each change in the independent variable. It could be the raw data, or some value obtained by processing the raw data.

Enthalpy change of combustion – the heat energy released when one mole of pure compound is combusted in excess oxygen under standard conditions.

Evaluated/Evaluation – an assessment of the reliability and precision of the raw data recorded during an experiment, the other limitations of the techniques, and the conclusions.

Evidence – data that has been validated.

Explanation/explain – give a detailed account including reasons or causes.

Extrapolate – to estimate the value of a variable outside the range used to establish a mathematical relationship between the variables, assuming that this relationship continues to be valid.

Fair test – a test in which only the independent variable has been allowed to affect the dependent variables.

Green chemistry – chemistry that aims to design products and processes that minimize the use and production of hazardous substances and wastes.

Hypothesis – a tentative explanation based on a scientific model of the observed chemical phenomenon you are investigating using the scientific method.

Independent variable – the variable that you systematically change (across a range) in an investigation.

Interpolate – to estimate a value for a variable between two or more known values. This is frequently done using a line graph.

Interval – the quantity between readings.

Investigation – a scientific study consisting of a controlled experiment in the chemistry laboratory.

Justify – using scientific ideas to explain data.

Key variable – an important variable with a large effect.

Literature value – a value from the chemical literature of a physical constant or experimental measurement.

Measurement error – the difference between a measured value and the literature value.

Measuring/measure – obtain a measured value for a quantity.

Methodology – the methods/ techniques (including statistical tests and controls) used to carry out an investigation.

Model – a representation that describes an explanation of the functioning, structure or relationships within a system or idea.

Percentage error – a percentage error is a random error expressed as a percentage of the value measured.

Plagiarism – the representation, intentionally or unwittingly, of the ideas, words or work of another person without proper, clear and explicit acknowledgment.

Precision – precise measurements are ones in which there is very little spread about the mean value.

Precipitation reaction – a reaction that involves the formation of an insoluble salt when two solutions containing soluble salts are combined. The insoluble salt formed is known as the precipitate.

Prediction – predictions are derived from a hypothesis and describe the results you expect to obtain from an investigation.

Primary data – raw data collected directly by a person from experiments.

Processed data – raw data that have been mathematically processed.

Processed variable – a variable that can be produced by transforming a measured variable through mathematical manipulation.

Propagation of errors – calculating the overall random error from a sequence of mathematical operations involving the processing of data containing random errors.

Qualitative data – observations not involving measurements; for example, colour changes during an experiment, the release of gases (colour and odour) and decrepitation (crackling noises) during heating.

Quantitative data – numerical data (with units and random uncertainty) from measurements, such as the pH during the determination of a titration curve (where volume of titrant is the independent variable).

Quantitative relationship – a relationship between variables that can be described by a mathematical equation.

Random error – these cause readings to be spread about the mean value, due to results varying in an unpredictable way from one measurement to the next. They result from limitations in the accuracy of the measuring scale used.

Range – the difference between the smallest and largest values. Usually applied to the independent variable.

Raw data – data that have not yet been processed or analyzed.

Regression – a calculation method (using least squares) to determine the best linear equation to describe a set of *x* and *y* data pairs.

Reliability – the results of an investigation may be considered reliable if the results can be repeated within experimental error.

Repeat – recording measurements twice or more to improve reliability, precision and accuracy.

Repeatability – precision obtained when measurement results are produced in one laboratory, by a single student, using the same equipment under the same conditions.

Replicate – repetition of the entire experiment run at the same time to record repeat measurements and observations.

Reproducible – a measurement is reproducible if the investigation is repeated by another person, or by using different equipment or techniques, and the same results are obtained.

Resolution – the smallest change in the quantity being measured that gives a measureable change in the reading from a particular measuring instrument. The smallest division of a scale that can be easily read.

Risk assessment – a consideration of the possible chemical and other hazards that could be present during an investigation as well as the environmental impact of disposal.

Scientific method – the use of controlled observations and measurements during an experiment to test a hypothesis.

Scientific notation – a method for expressing a given quantity as a number having significant digits necessary for a specified degree of accuracy, multiplied by 10 to the appropriate power.

Secondary data – data collected by a person or group other than the person or group using the data.

Sensitivity – the smallest change that can be detected by a measurement by an instrument.

Serial dilution – any dilution where the concentration decreases by the same factor in each successive step.

Significant figures – the digits of a number that are used to express it to the required degree of accuracy, starting from the first non-zero digit.

Simulation – a representation of a process or system that imitates a real or idealized situation.

Standard deviation – a measurement of the spread of normally distributed data around a mean.

Standard solution – a solution with an accurately known concentration, prepared from a primary standard. This is a substance dissolved in a known volume of water to give a standard solution.

Systematic error – these cause readings to be in error by a consistent amount each time a measurement is made. Sources of systematic error can include the environment, methods of observation or instruments used. They are not reduced by repetition.

Tests – investigations involve a number of tests where one variable is manipulated or changed.

Titrant – the substance added during the titration.

Trend – the general relationship shown by a set of related measurements.

Trial run – a trial run can be used to find the scale, range and number of variables and help in choosing apparatus and instruments.

Uncertainty – the range that will contain the true value of the measurement.

Valid conclusion – a conclusion supported by valid data, obtained from an appropriate experimental design and based on chemical reasoning.

Validity – a measure, including statistical, of the confidence in a conclusion. It depends on the range and reliability of observations and measurements.

Validity (of methodology) – suitability of the investigative procedure to answer the research question.

Variable – a factor that is being changed, measured, or kept the same during an investigation.

Zero error – any indication that a measuring system gives a false reading rather than the true value of a measured quantity.

Index

Answers

Chapter 1

Page 4

1 You can check its density against air ($1.225\,\text{kg}\,\text{m}^{-3}$ or $0.001\,225\,\text{g}\,\text{cm}^{-3}$ at sea level at a temperature of $15\,°C$). Densities of gases at specified temperatures can be found in data books and websites. Alternatively, you can fill a balloon with the gas and observe whether it floats or sinks in air.

Downward displacement of water is not an ideal method for collecting carbon dioxide because carbon dioxide has some solubility in water. It dissolves and reacts reversibly to form carbonic acid, which ionizes.

$$CO_2(g) + H_2O(l) \rightarrow H_2CO_3(aq) \rightleftharpoons H^+(aq) + HCO_3^-(aq)$$

Page 7

2 $0.079\,\text{g}$

3 $98970\,\text{Pa}$

4 $0.001\,20\,\text{mol}$

5 $65.83\,\text{g}\,\text{mol}^{-1}$

6 $+13.23\,\%$

7 The butane gas is not appreciably soluble in water and does not react with water. The gas collected in the eudiometer and the surrounding water are in thermal equilibrium.

8 Capturing gas from the lighter in the eudiometer; removing water that collected in the butane lighter before the final mass was recorded; water possibly getting into the lighter while the butane gas was being expelled.

Page 8

9 Temperature of the solutions; immersion depth of the electrodes; distance between the two electrodes; the nature of the electrolyte in the salt bridge and its dimensions.

10 Temperature and concentration of the cations (electrolytes).

11 Cell voltages added to table.

$[Mg^{2+}(aq)]/\text{mol}\,\text{dm}^{-3}$	1.000	0.100	0.010	0.001	1.000	1.000	1.000
$[Cu^{2+}(aq)]/\text{mol}\,\text{dm}^{-3}$	1.000	1.000	1.000	1.000	0.100	0.01	0.001
Cell voltage/V	2.710	2.740	2.769	2.799	2.680	2.651	2.621

12 Amount (in moles) of copper atoms $= \dfrac{0.030\,\text{g}}{63.55\,\text{g}\,\text{mol}^{-1}} = 4.72 \times 10^{-4}\,\text{mol}$

13 Amount (in moles) of copper atoms $= \dfrac{It}{2F}$

$F = \dfrac{It}{2} \times$ amount of copper atoms

$F = \dfrac{0.30 \times 300.0}{2} \times 4.72 \times 10^{-4}$

$F = 95\,338\,\text{C} \approx 95\,300\,\text{C}\,\text{mol}^{-1}$ (3 sf)

14 Absolute random uncertainty for the Faraday constant

$$= \left(\frac{0.003}{0.030} + \frac{0.2}{300.0} + \frac{0.02}{0.30} \right) \times 95\,338\,C\,mol^{-1}$$

$$= 15\,953\,C\,mol^{-1}$$

$$= 16\,000\,C\,mol^{-1} \text{ (to 2 sf)}$$

15 Faraday constant: $(95\,338 \pm 13\,000)\,C\,mol^{-1}$

$$\% \text{ error} = \frac{|+95\,338 - (+96\,485)|}{96\,485} \times 100\,\% = 1.1889\,\% = 2\,\% \text{ (to 1 sf)}$$

■ Page 10

16 Ethanoic acid + ethanoate ions = 0.2 M

Using the Henderson–Hasselbalch equation:

$$pH = pK_a + log_{10} \left\{ \frac{[ethanoate]}{[ethanoic\ acid]} \right\}$$

$$5.0 = 4.76 + log_{10} \left\{ \frac{[ethanoate]}{[ethanoic\ acid]} \right\}$$

$$\left\{ \frac{[ethanoate\ ions]}{[ethanoic\ acid]} \right\} = \frac{1.74}{1}, \text{ so [ethanoate] as a fraction of total}$$

(ethanoic acid + ethanoate ions) $= \frac{1.74}{2.74} = 0.635$

Therefore, [ethanoate ions] $= 0.635 \times 0.2\,M$ total $= 0.127\,M$; and [ethanoic acid] $= 0.073\,M$.

To prepare the buffer:

- Take $800\,cm^3$ of water and add $0.20\,mol$ $(12.0\,g)$ of ethanoic acid.

- Add $127\,cm^3$ of $1.00\,M$ NaOH to produce $0.127\,M$ sodium ethanoate.

- Add water to make up to $1000\,cm^3$.

17 The maximum mass of NH_4Cl that can dissolve in $100\,cm^3$ of water

$$= \frac{6.95}{1000} \times 100 \times 53.5 = 37.2\,g$$

Assuming a temperature change of $5\,°C$ to be measured and no heat loss to surroundings: amount of $NH_4Cl \times 15\,000 = 100 \times 4.18 \times 5$

Therefore, amount of $NH_4Cl = 0.139\,mol$, so minimum mass to use $= 0.139\,mol \times 53.5\,g\,mol^{-1} = 7.45\,g$

18

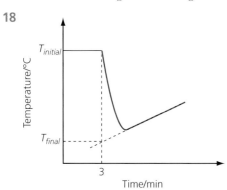

Change in temperature with time

■ Page 11

19 A copper can is preferred to a glass beaker because copper is an excellent thermal conductor and will transfer more heat than a glass beaker into the water.

20 Since the base of the copper calorimeter comes into contact with the heat source directly, the temperature will be significantly higher than that of the water.

▥ Page 12

21 The standard enthalpy change of combustion is the heat released when 1 mole of substance in its standard state is completely burned in oxygen under standard conditions (298 K and 1 atm pressure).

$$C_6H_8(l) + 8O_2(g) \rightarrow 6CO_2(g) + 4H_2O(l)$$

22 Measure $100\,cm^3$ of water into a calorimeter (using a $100\,cm^3$ measuring cylinder). Record the initial temperature of the water, T_1 (using a thermometer). Fill the spirit burner with cyclohexa-1,3-diene and weigh it. Record the mass as m_1. Heat the water until a temperature rise of about $10\,°C$ has been obtained. Extinguish the flame but continue stirring. Record the maximum temperature, T_2, reached. Allow the apparatus to cool and reweigh the spirit burner and its contents. Record the mass as m_2.

23 Amount of cyclohexa-1,3-diene burned $= \dfrac{(m_1 - m_2)}{80.0}\,mol$

Thermal energy absorbed by the water $= 100 \times 4.18 \times (T_2 - T_1)\,J$

Enthalpy of combustion $= \dfrac{\text{thermal energy absorbed by water (in calorimeter)}}{\text{amount of cyclohexa-1,3-diene burned}}$

24 Assume that the calorimeter has a negligible heat capacity. ΔH_c calculated will be numerically lower / less exothermic than expected (as some heat is absorbed by the calorimeter). Assume complete combustion of the hydrocarbon has taken place. ΔH_c calculated will be numerically lower / less exothermic than expected (as some heat will be released).

▥ Page 14

25 A dilatometer is an instrument comprising a glass vessel that measures small volume changes caused by a physical change or chemical reaction, for example, polymerization.

26 Advantages: only one reaction mixture needs to be prepared; continuous measurements can be recorded that may be achieved via data-logging; no disruption of the reaction by withdrawing samples for analysis.

Disadvantages: expensive equipment may be needed; not suitable for all reactions; raw data may need to be processed to convert to concentrations.

27 Advantages: simple to carry out; can use readily available and cheap laboratory apparatus; the raw data is relatively easy to process.

Disadvantages: only averaged rates can be measured, which can be approximated to initial rates only if the time measured is relatively short; a number of reaction mixtures need to be prepared.

▥ Page 15

28 First order with respect to both iodide and peroxydisulfate ions.

$$\text{rate} = k(0.038\,mol\,dm^{-3}) \times (0.060\,mol\,dm^{-3}) = 1.5 \times 10^{-5}\,mol\,dm^{-3}\,s^{-1}$$

29 $k = \dfrac{1.5 \times 10^{-5}\,mol\,dm^{-3}\,s^{-1}}{(0.038\,mol\,dm^{-3})(0.060\,mol\,dm^{-3})} = 6.6\,mol^{-1}\,dm^3\,s^{-1}$

When $[S_2O_8^{2-}] = [I^-] = 0.050\,mol\,dm^{-3}$;

$$\text{rate} = (6.6 \times 10^{-3}\,mol^{-1}\,dm^3\,s^{-1}) \times (0.050\,mol\,dm^{-3}) \times (0.050\,mol\,dm^{-3})$$

$$= 1.6 \times 10^{-5}\,mol\,dm^{-3}\,s^{-1}$$

▥ Page 16

30 Amount of magnesium $= \dfrac{0.125\,g}{24.31\,g\,mol^{-1}} = 5.14 \times 10^{-3}\,mol$

Amount of hydrochloric acid = $\dfrac{15.00}{1000}\,dm^3 \times 0.500\,mol\,dm^{-3} = 7.5 \times 10^{-3}\,mol$

Only $\dfrac{7.5 \times 10^{-3}\,mol}{2} = 3.75 \times 10^{-3}\,mol$ of magnesium will be reacted, hence the magnesium is present in excess and the hydrochloric acid is the limiting reagent.

Maximum volume of hydrogen = $\dfrac{7.5 \times 10^{-3}\,mol}{2} \times 22.7\,dm^3\,mol^{-1} = 85\,cm^3$

31 The magnesium oxide layer needs to be removed to expose the underlying magnesium.

32 Tangents to the curve should be drawn on the graph of total volume of gas against time. A linear graph of rate against time confirms first order.

33 No naked flames – hydrogen gas is flammable

34 Completed table:

Volume of 0.40 mol dm^{-3} sodium thiosulfate/ cm^3 ± 0.05 cm^3	Volume of 2.00 mol dm^{-3} hydrochloric acid/ cm^3 ± 0.05 cm^3	Volume of water/ cm^3 ± 0.05 cm^3	Time/s
40.00	30.00	0.00	–
30.00	20.00	20.00	–
40.00	15.00	15.00	–
40.00	5.00	25.00	–
30.00	10.00	30.00	–

35 So that the total volume of the reaction mixture remains constant for the different sets of experiments and hence the volume of reagent (thiosulfate ions) used is proportional to its concentration in the reaction mixture.

36 The volume of sodium thiosulfate(VI) used is directly proportional to its concentration since the total volume of reaction mixture is kept constant.

37 The relative rate of reaction should be inversely proportional to the time taken for the cross under the flask to be obscured since the amount of sulfur produced in every set of experiments is similar.

38 Hydrogen sulfide, H_2S. It turns moist lead(II) ethanoate paper black.

■ Page 18

39 Phenolphthalein would be a suitable indicator. Methyl orange would not be suitable for the titration of a weak acid and a strong base.

40 The flasks need to be securely stoppered when they are left for a week. If this is not done, some of the mixture could evaporate. Shaking was done to ensure solutions were homogeneous and reactants were physically mixed.

41 The density of ethanol is needed, since mass = volume × density.

42 The potassium hydroxide solution titre for flask 4 seems to be anomalous.

43 A lower value for K_c would be expected if the reaction had not reached equilibrium.

Chapter 2

■ Page 19

1 Sodium hydroxide is not a primary standard because it absorbs moisture from the air (deliquescesces) and dissolves in it to form a very concentrated solution. In addition, both solid sodium hydroxide and a solution of it react with carbon dioxide from the air to form sodium carbonate. Consequently, it is unstable in moist air and so does not meet the requirements of a primary standard.

Page 20

2 The results of the titration will be unaffected. Although the solution will be diluted, the amount (in moles) of analyte will be unchanged and this is the critical factor.

Page 22

3 In the titration of an alkaline mixture, for example, NaOH and $NaHCO_3$ (or Na_2CO_3 and $NaHCO_3$), with strong acid, two indicators are used: phenolphthalein and methyl orange. Phenolphthalein is a weak organic acid and gives an end-point between pH 8 and 10, while methyl orange is a weak base and indicates an end-point sharply between pH 3.1 and 4.4.

Mixture to be estimated	Maximum pH change during reaction	First indicator	Second indicator
NaOH and $NaHCO_3$		Phenolphthalein	
$NaHCO_3$/HCl	14–1	First titre estimates NaOH	
$NaHCO_3$/HCl	8–1		Methyl orange Second titre estimates $NaHCO_3$
Na_2CO_3 and $NaHCO_3$			
Na_2CO_3/HCl	13–8 Na_2CO_3 is converted to $NaHCO_3$	Phenolphthalein; first titre	
$NaHCO_3$/HCl (both the $NaHCO_3$ from above and the $NaHCO_3$ originally present in the mixture)	8–10		Methyl orange Second titre value

4 The end-point is when there is a colour change in the indicator or chemical species; the equivalence point occurs when there are stoichiometric quantities of reacting chemicals.

Page 24

5 $\dfrac{2.2\,\text{mg}}{0.25\,\text{dm}^3} = 8.8\,\text{mg}\,\text{dm}^{-3} = 8.8\,\text{ppm}$

6 Mass of ethanoic acid in $10\,\text{cm}^3$ of white vinegar solution

$= \dfrac{0.8393\,\text{mol} \times 60.06\,\text{g}\,\text{mol}^{-1}}{100} = 0.5040\,\text{g}$

Mass of white vinegar $= 10\,\text{cm}^3 \times \dfrac{1.006\,\text{g}}{1\,\text{cm}^3} = 10.06\,\text{g}$

Mass % $= \dfrac{0.5040\,\text{g}}{10.06\,\text{g}} \times 100 = 5.011\,\%$

7 Solutions of hydrogen peroxide are often sold according to their 'volume strength'. The volume strength of a solution of aqueous hydrogen peroxide is measured by the number of volumes of oxygen released when it is completely decomposed under standard conditions (0 °C, 1 atmosphere pressure).

For example, 20 volume strength hydrogen peroxide solution means $1\,\text{cm}^3$ of the solution will release $20\,\text{cm}^3$ of oxygen gas (when completely decomposed). Volume strengths can be converted to other measures of concentration, for example, molarity ($\text{mol}\,\text{dm}^{-3}$) and percentage by volume.

8 Amount of $Cr_2O_7^{2-} = 0.0100 \times \left(\dfrac{14.00}{1000}\right) = 1.4 \times 10^{-4}\,\text{mol}$

Amount of Fe^{2+} in $25.00\,\text{cm}^3 = 1.4 \times 10^{-4}\,\text{mol} \times 6 = 8.4 \times 10^{-4}\,\text{mol}$

Amount of Fe^{2+} in $250\,\text{cm}^3 = 8.4 \times 10^{-3}\,\text{mol}$

Mass of iron(II) sulfate in iron tablet $= 8.4 \times 10^{-3}\,\text{mol} \times 151.92\,\text{g}\,\text{mol}^{-1} = 1.28\,\text{g}$

9 ■ Filter the mixture using a filter funnel to remove the excess zinc and collect the filtrate containing iron(II) ions.

   ■ Pipette 25.00 cm³ of the iron(II) ion solution (filtrate) into a conical flask and add 1 cm³ of N-phenylanthranilic acid using a 10 cm³ measuring cylinder / graduated dropper. Fill a burette with the standard potassium dichromate(VI) solution and record the initial reading on the burette.

   ■ Titrate the iron(II) solution in the conical flask until the solution turns (from pale green to yellow) to violet; record the final reading on the burette.

   ■ Repeat the titrations and average consistent results.

10 Add a known mass of excess zinc powder to the solution containing a mixture of iron(II) and iron(III) ions. Filter the resultant solution, recover and dry the residue containing the unreacted zinc powder. The difference in mass of zinc can be used to calculate the amount (mol) of iron(III) ions present. Add zinc powder to the beaker containing the 250 cm³ solution of iron(II) and iron(III) ions with stirring until no more zinc reacts and dissolves.

■ Page 26

11 ■ Weigh accurately an empty dry boiling tube.

   ■ Add about 3.00 g of the sample and weigh the boiling tube and its contents.

   ■ Heat the sample in the boiling tube strongly with the Bunsen burner flame for about 5 minutes.

   ■ Allow the boiling tube and contents to cool on a heat-proof mat.

   ■ When it is cool reweigh the boiling tube and its contents.

   ■ Repeat the heating, cooling and reweighing until there is no further loss in mass (that is, the final mass within 0.05 g). This is to ensure that all the magnesium carbonate has decomposed.

   ■ A suitable results table is shown below.

Mass of boiling tube and sample/g	B
Mass of empty boiling tube/g	A
Mass of sample/g	B – A
Mass of boiling tube and contents (after heating to constant mass)/g	C
Final mass of residue/g	C – A
Loss in mass/g	B – C

■ Page 27

12

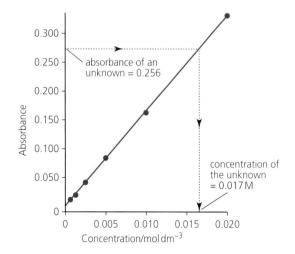

Chapter 3

▓ Page 31

1 Before the liquid starts boiling, the temperature in the column is much lower than that in the liquid mixture, as there is no vapour to heat up the thermometer. However, when the liquid starts boiling and its vapour rises, the temperature of the vapour is similar to that of the boiling liquid. The temperature of a substance remains constant while it is boiling, so when the thermometer shows a constant temperature, you are collecting the desired organic product. By placing the thermometer at the point where the vapour exits into the Liebig condenser, you know that you are measuring the temperature of the vapour being collected. The condensation point is the same numerical value as the boiling point of a liquid.

2 Letting the cooling water enter the condenser at the bottom ensures that the temperature is lowest at this end. This means all the vapour is condensed without any loss.

▓ Page 32

3 Oiling out occurs when a compound is very impure or when it has a melting point lower than the boiling point of the solvent. Instead of forming crystals, an oily liquid is formed. The oil should be redissolved by heating the solution and cooling the dilute solution very slowly.

▓ Page 33

4 Molar mass of cyclohexanol = $[(6 \times 12.01) + 11.11 + 16.00 + 1.01] = 100.18\,\mathrm{g\,mol^{-1}}$

Mass of cyclohexanol = $6.818\,\mathrm{cm^3} \times 0.811\,\mathrm{g\,cm^{-3}} = 6.000\,\mathrm{g}$

Amount of cyclohexanol used = $\dfrac{6.000\,\mathrm{g}}{100.18\,\mathrm{g\,mol^{-1}}} = 0.06\,\mathrm{mol}$

5 Molar mass of cyclohexene = $(6 \times 12.01)\,\mathrm{g\,mol^{-1}} + (1.01 \times 10)\,\mathrm{g\,mol^{-1}}$

$$= 82.06\,\mathrm{g\,mol^{-1}}$$

Amount of cyclohexene that should be formed (theoretical yield)
= $0.06\,\mathrm{mol} \times 82.06\,\mathrm{g\,mol^{-1}} = 4.92\,\mathrm{g}$

6 Percentage yield = $\dfrac{1.85\,\mathrm{g}}{4.92\,\mathrm{g}} \times 100 = 37.6\,\%$

7 The boiling point of cyclohexene is the lower of the two so it distills off first. The boiling point of cyclohexanol is higher so it remains in the flask.

8 The carbon (soot) must come from the carbon atoms of the cyclohexanol molecules, so this carbon is not converted to cyclohexene.

9 The mixture should be added to cold water. Gloves, safety glasses and laboratory coat must be worn.

10 See table below.

Reagent(s)	Observation
Bromine water or bromine	Yellow brown or orange solution becomes colourless / is decolorized
Potassium manganate(VII) and dilute sulfuric acid solution	Purple solution becomes colourless / is decolorized
Potassium manganate(VII) and sodium carbonate / sodium hydroxide solution	Purple solution forms a brown precipitate

▓ Page 34

11 The product is impure and needs to be purified by recrystallization; the compound is not pure N-ethylethanamide; the thermometer used should be calibrated.

12 The method of mixed melting points involves mixing a pure sample of the organic compound you have prepared with a pure commercial sample of the same organic compound. Equal masses of the two compounds are ground together and the melting point of the mixture is then measured. If the measured melting point is sharp and close to the expected value, then the two organic compounds *are* identical. However, if the two compounds are not the same then the melting point of the mixture will be lower and the melting range broader, because each substance will act as an impurity on the other.

13 Brady's reagent is acidified 2,4-dinitrophenylhydrazine. This reacts with a carbonyl compound to form an orange-coloured crystalline precipitate, which is the solid derivative. The derivative is then recrystallized and its melting point determined. Each carbonyl derivative has a unique melting point.

Chapter 4

■ Page 36

1 Scientific notation: 1.002×10^3, 5.4×10^1, 6.9263×10^9, -3.93×10^2, 3.61×10^{-3}, 3.8×10^{-3}

2 Ordinary notation: 1930, 30.52, −429, 6 261 000, 0.000 000 095 13

■ Page 37

3 14.44 has four significant figures. All non-zero digits are significant.

9000 has no decimal point, so the zeros may or may not be significant. We cannot tell how many significant figures there are. When considering numbers with zeroes at the end we must state the number of significant figures.

3000.0 has five significant figures. The decimal point implies that we have measured to the nearest 0.1.

1.046 has four significant figures. Zeros between digits are significant.

0.26 has two significant figures. Zeros to the left of the decimal point only fix the position of the decimal point. They are not significant.

6.02×10^{23} has three significant figures. The rules are the same when dealing with numbers expressed in standard form so 6×10^{23} has one significant figure, but 6.02×10^{23} has three significant figures.

4 654.389 becomes 654 because the first non-significant digit is 3.

65.4389 becomes 65.4.

654 389 becomes 654 000 because we need to put zeros in to hold the place values.

56.7688 becomes 56.8 because the first non-significant digit is 6.

0.035 422 10 becomes 0.0354. Note that three significant figures is not the same as three decimal places, which would give 0.035.

■ Page 38

5 26.3

6 1.2×10^3 (1200)

7 0.5

8 4.4

9 3.0

10 35.73

11 2.73

12 3.0771

Page 41

13 Half of the smallest division on the scale implies a reading of −2.5 °C with an uncertainty of ±0.5 °C. If the scale is interpolated with a magnifying glass, then the uncertainty may be close to ±0.2 °C.

Disadvantages of alcohol: it cannot measure high temperatures because of its relatively low boiling point; it wets the inside of the tube; it is colourless (so must be dyed to make it visible) and liable to evaporate.

Advantages of mercury: it is visible because of its silvery lustre; its thermal expansion is more regular than that of alcohol; it does not wet the inside of the tube and it can measure temperatures more quickly and precisely than alcohol.

Page 42

14 Percentage error $= \left| \dfrac{39.7 - 44.2}{44.2} \right| \times 100 = 10.2\,\%$ error

Page 43

15 **Calculate the amount (in moles) of Fe^{2+} ions in 20.0 cm³ of the solution**

Volume of $MnO_4^- = \dfrac{(15.20 + 15.30 + 15.20)}{3}$

$\qquad = 15.23\,cm^3 \pm 0.10\,cm^3$

$\qquad = 15.23\,cm^3 \pm 0.657\,\%$

Amount of $MnO_4^- = 0.015\,23 \times 0.0500$

$\qquad = 7.615 \times 10^{-4}\,mol$

Amount of Fe^{2+} in 20.0 cm³ $= 7.615 \times 10^{-4} \times 5$

$\qquad = 3.8075 \times 10^{-3}$

$\qquad = 3.81 \times 10^{-3} \pm 1.7\,\%$

Uncertainty $= \pm\, \dfrac{0.10}{15.23} \times 100\,\% = \pm\, 0.657\,\%$

$\qquad = \pm \left(0.657 + \dfrac{0.0005}{0.0500} \times 100\,\% \right)$

$\qquad = \pm\, 1.657\,\%$

Calculate the concentration of Fe^{2+} ions in the solution

Concentration of $Fe^{2+} = \dfrac{(3.8075 \times 10^{-3})}{0.02000}$

$\qquad = 0.1904\,mol\,dm^{-3}$

$\qquad = 0.190\,mol\,dm^{-3} \pm 1.8\,\%$

Uncertainty $= \pm \left(1.657 + \dfrac{0.03}{20.00} \times 100\,\% \right)$

$\qquad = 1.807\,\%$

Calculate the mass of iron in the steel

Amount of moles Fe^{2+} = number of moles of Fe

$\qquad = 0.1904\,mol\,dm^{-3} \times 0.250\,00\,dm^3$

$\qquad = 0.0476\,mol$

Mass of iron $= 0.0476\,mol \times 55.85\,g\,mol^{-1}$

$\qquad = 2.658\,g$

$\qquad = 2.66\,g \pm 1.8\,\%$

Uncertainty $= \pm \left(1.807 + \dfrac{0.03}{250.00} \times 100\,\% \right)$

$\qquad = 1.819\,\%$

Determine the percentage composition by mass of iron in the steel sample

Percentage composition of iron in steel

$= \dfrac{2.658}{2.923} \times 100\,\%$

$= 90.93\,\%$

$= 90.9\,\% \pm 2\,\%$ (relative)

$= 90.9 \pm 1.8\,\%$ (absolute)

Uncertainty $= \pm \left(1.927 + \dfrac{0.002}{2.923} \times 100\,\% \right)$

$\qquad = 1.995\,\%$

Page 44

16 Mass of carbon dioxide = (17.46 g − 16.61 g) = 0.85 g

17 Amount of carbon dioxide produced $= \dfrac{mass}{molar\ mass} = \dfrac{0.85\,g}{44.01\,g\,mol^{-1}} = 0.0193\,mol$

18 Mass of MCO_3 = (17.46 g − 15.23 g) = 2.23 g

19 Amount (mol) of MCO_3 = amount (moles) of CO_2 = 0.0193 mol

Molar mass of $MCO_3 = \dfrac{mass}{molar\ mass} = \dfrac{2.23\,g}{0.0193\,mol} = 116\,g\,mol^{-1}$

Molar mass of M = 116 − [12.01 + (3 × 16.00)] = 56.0 g mol⁻¹

20 Mass of CO_2 = 0.85 g ± 0.02 g

The amount (mol) of CO_2 lies between $\dfrac{0.83\,g}{44.01\,g\,mol^{-1}} = 0.0189\,mol$ and

$\dfrac{0.87\,g}{44.01\,g\,mol^{-1}} = 0.0198\,mol$

Since the ratio of MCO_3 to CO_2 is 1 : 1 the amount (mol) of MCO_3 also lies between 0.0189 and 0.0198 mol.

Mass of MCO_3 = 2.23 g ± 0.02 g

Maximum molar mass of $MCO_3 = \dfrac{largest\ mass}{smallest\ amount} = \dfrac{2.25\,g}{0.0189\,mol} = 119\,g\,mol^{-1}$

Therefore, the maximum molar mass of M = 119 g mol⁻¹ − 60.0 g mol⁻¹ = 59 g mol⁻¹

Minimum molar mass of $MCO_3 = \dfrac{smallest\ mass}{largest\ amount} = \dfrac{2.21\,g}{0.0198\,mol} = 112\,g\,mol^{-1}$

Therefore, the minimum molar mass of M = 112 g mol⁻¹ − 60.0 g mol⁻¹

= 52 g mol⁻¹

The random uncertainty of the molar mass of M is that it is between 52 and 59 g mol⁻¹.

Page 45

21 It would lead to a systematic error since the student's titre volumes would *all* be greater than the true value or would *all* be less than the true value.

22 There would be a systematic error in the measured pH values since *all* of them would be less than the true values by about 0.10 pH units.

23 Calculated $M_m(HA)$ too low:

M(NaOH) → V(NaOH) → n(NaOH) → n(HA) → M_m(HA)

(M = n ÷ V) and (M_m = m ÷ n)

24 Calculated $M_m(HA)$ not affected. Water does not change n(HA); it changes only M(HA) via dilution, it is not a reactant.

25 Calculated $M_m(HA)$ too high:

equivalence point → n(NaOH) → n(HA) → M_m(HA)

(expected pH higher)

Page 48

26 Average = 0.105; $S^2 = 1.4 \times 10^{-7}$

To be discarded $|Z_{calc}| \geq Z_{0.025}$; $|Z_{calc}| \geq 1.96$

$\dfrac{0.105 - 0.1021}{\sqrt{1.4\times10^{-7}}\,/\,\sqrt{5}} = 3.586$

Since $|Z_{calc}| \geq 1.96$, then discard 0.1021

Chapter 5

▨ Page 53

1   A potentiometric titration involves the measurement of the potential (voltage) of
a suitable indicator electrode with respect to a reference electrode as a function of
titrant volume. Potentiometric titrations provide more reliable data than data from
titrations that use chemical indicators and are particularly useful with coloured or
turbid (cloudy) solutions and for detecting the presence of unsuspected species.

A typical set up for potentiometric titrations is shown in Figure A3. Titration
involves measuring and recording the cell potential (in units of millivolts or
pH) after each addition of titrant. The titrant is added in large amounts at the
start and in smaller and smaller amounts as the end-point is approached (as
indicated by larger changes in response per unit volume). Sufficient time must be
allowed for the attainment of equilibrium after each addition of the reagent by
continuous stirring. For this a magnetic stirrer with a stirring magnet bar is used.

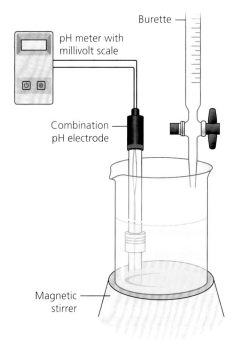

Apparatus for a potentiometric titration

Chapter 7

▨ Page 65

1   ■   How does the rate of hydrolysis of urea depend on the concentration of
enzyme and substrate?

■   What is the optimum pH for the reaction between urea and urease?

■   Does the optimum pH depend on urea concentration?

■   Does the effect of the urea concentration depend on the pH?

■   What is the optimum temperature for the reaction?

■   Does this depend on the pH or the urea concentration?

■   How is the reaction rate affected by the presence of other substances, such
as copper(II) or lead(II) ions?

2 A method is not specified, for example, measurement of turbidity or pH, nor is the identity of the alkali. An independent variable, for example, ethanal concentration, and controlled variables, such as temperature and alkali concentration, are not specified. The aim is not specified, for example, establishing the rate, individual orders or overall orders, determining the activation energy, identifying intermediates, and so on.

Chapter 9

■ Page 78

1 **Quantified weaknesses and limitations**

The precision of the balance was limited to ±0.01 g. Using an analytical balance (±0.001 g) will reduce the percentage random uncertainty by a factor of 10. The balance will be tested with known masses against two other similar analytical balances to ensure no zero error is present.

The copper powder may not have been thoroughly dried and should be washed with cold water to remove soluble salts. Dry the copper powder to constant mass at a higher temperature and for a longer period of time (until constant mass) to ensure that all the water has been removed by evaporation.

Unquantified weaknesses and limitations

The surface of the iron particles will have hydrated iron(III) oxide, rust. The amount of reacting iron will be less, which will reduce the amount of copper formed. Fresh iron powder of higher purity (stored in an inert atmosphere or vacuum packed) is to be used.

Copper(II) ions are acidic in aqueous solution:
$[Cu(H_2O)_6]^{2+}(aq) \rightarrow [Cu(H_2O)_5(OH)]^+(aq) + H^+(aq)$.
Some iron may be involved in a side reaction:
$Fe(s) + 2H^+(aq) \rightarrow Fe^{2+}(aq) + H_2(g)$.
This is a systematic error since it reduces the amount of iron involved in the main reaction.

It is assumed that there is no reaction between the reagents and the crucible. This may not be true with hot chemicals and any side reactions will introduce a very small systematic error. An inert platinum crucible could be used.

It is assumed that the iron powder is sufficiently fine (as well as pure) that it is completely displaced by copper. The displacement reaction should be repeated with purer iron powder of smaller particle size (greater surface area) with hot copper(II) sulfate solution. It is assumed that there is sufficient time for the reaction to go to completion.

Integer atomic masses were used in the calculation, limiting the precision of the calculation. Relative atomic masses with two decimal places should be used to improve the accuracy of the calculation.

The copper(II) sulfate solution was prepared by the school's technician. The concentration will have an unknown associated random uncertainty. You should make up your own solution and quantify the error involved or ask the technician what apparatus was used so you can quantify the overall random error.